Springer Polar Sciences

Series editor
James Ford, Priestley International Centre for Climate, University of Leeds, Leeds, United Kingdom

Springer Polar Sciences

Springer Polar Sciences is an interdisciplinary book series that is dedicated to research on the Arctic and sub-Arctic regions and Antarctic. The series aims to present a broad platform that will include both the sciences and humanities and to facilitate exchange of knowledge between the various polar science communities.

Topics and perspectives will be broad and will include but not be limited to climate change impacts, environmental change, polar ecology, governance, health, economics, indigenous populations, tourism and resource extraction activities.

Books published in the series will have ready appeal to scientists, students and policy makers.

More information about this series at http://www.springer.com/series/15180

Minori Takahashi
Editor

The Influence of Sub-state Actors on National Security

Using Military Bases to Forge Autonomy

Springer

Editor
Minori Takahashi
Slavic-Eurasian Research Center and Arctic Research Center
Hokkaido University
Hokkaido, Japan

ISSN 2510-0475 ISSN 2510-0483 (electronic)
Springer Polar Sciences
ISBN 978-3-030-01676-0 ISBN 978-3-030-01677-7 (eBook)
https://doi.org/10.1007/978-3-030-01677-7

Library of Congress Control Number: 2018962927

This Springer imprint is published by the registered company Springer Nature Switzerland AG
The registered company address is: Gewerbestrasse 11, 6330 Cham, Switzerland

Preface

I have been studying and doing research driven by my interest in international politics regarding the Arctic, including Greenland. When I was a graduate student in the latter half of the 2000s and the first half of the 2010s, through my fieldwork and studies at the University of Greenland/Ilisimatusarfik I built a human network consisting of many knowledgeable persons, among whom were not only researchers but also citizens, professional groups, officials from government agencies, and corporations, and fostered the budding ideas for my research. However, when I returned to my home country of Japan, rather than spending time on research itself, I found myself troubled (especially when in the circles of the Japan Association of International Relations) by the need to think of how to create a "cooking recipe" that would allow others "to consume the beast," so to speak. That is, I was repeatedly told by my academic mentor that since nobody would touch peculiar foods such as Greenland politics or Arctic politics if served raw, I had to think of a way to make them look tasty. At a meeting on international politics held in Japan, I was even asked "Is there such a thing as politics in the Arctic?," and as a person whose research is about Greenland and the Arctic politics, I have to say I experienced no shortage of such episodes. At least in Japan, where I am based, such was, overall, the reality until very recently.

However, since 2015 the environment has changed. In April of that year, one of the most authoritative international conferences on the Arctic region, the "Arctic Science Summit Week," was held for the first time in Japan at the initiative of the Science Council of Japan. And in September, a project entitled the "Arctic Challenge for Sustainability (ArCS)," the core of which is formed by the National Institute of Polar Research, Japan Agency for Marine-Earth Science and Technology, and Hokkaido University, commenced work as the first national flagship project on the Arctic bringing together social and natural sciences. Furthermore, in October of the same year, based on coordination with the ArCS, Japan formulated its first national policy regarding the Arctic, announcing to the domestic and international public its interest in the region. The formulation of the Arctic policy meant that Japan, albeit late, entered that area. In connection with these developments, at Hokkaido University, a state university in north Japan, a national center for Arctic research was founded, and in April of 2016, the Hub for the Promotion of Joint Research on Arctic (J-ARC Net), headquartered at that university and expected to function as a platform encouraging research, was established. Thus, the environment surrounding myself and my colleagues is now dramatically changing. Currently, in Japan a system dedicated to engaging the following versatile tasks related to the Arctic region is finally, little by little, taking shape: (1) offering recommendations regarding global environmental issues, (2) making contributions toward sustaining the livelihood and the cultural foundations of Arctic indigenous peoples, (3) providing scientific explanations of the mechanisms behind the climate change, (4) acting based on the rule of law and creating international networks, (5) exploring the possibility of a stable utilization of shipping lanes in the Arctic Ocean, (6) exploring the possibilities for developing natural resources in the Arctic region with an eye to diversifying own supply sources, and (7) analyzing the factors behind the changes in the security situation in the Arctic.

Of course, these activities are not meant to be limited only to the boundaries of Japan. For example, the government report "Achieving Innovative Solutions for Arctic Challenges," published on March 7, 2018, by the Advisory Board for the Promotion of Science and Technology Diplomacy, advocates the need to collaborate with other countries in the international community in order to leave an open and sustainable Arctic to the posterity and calls on Japan to play an active role in the appropriate fields while taking five Is, international, interdisciplinary, inclusive, ingenuity, and innovation, as the guiding principles. Although this book is not directly related to these developments, it does involve a desire to make a contribution in the same direction.

The outline of this book was directly shaped by the joint research project "Study of Security in the Arctic from the Standpoint of International Relations," supported by the 45th Research Grant for Humanities given by the Mitsubishi Foundation (principal investigator: Minori Takahashi). That was a joint effort started by three persons: Kousuke Saitou, who specializes in the US security policy; Shinji Kawana, who specializes in the politics of military bases; and myself. I was extremely fortunate to have been able to engage in such a project with these up-and-coming researchers. I still remember clearly the stimulating meeting in August 2015 in the Tokyo district of Hanzomon, when incipient ideas regarding our research were exchanged and the first step was made. Later, through their introduction, Yu Koizumi, who specializes in military issues and security policies in Russia and other former Soviet republics; Shino Hateruma, who studies US military bases in Okinawa; and Ayae Shimizu, who conducts research on the national security of the Philippines, all energetic researchers belonging to the same generation and abounding in personal charm, joined us, thanks to which our project received a strong boost.

Along the way we had the privilege to receive three grants from the Japan Arctic Research Network Center (J-ARC Net) Program for the Promotion of Joint Research for the following topics: "American Policy towards the Arctic and the Reaction of the Concerned Countries: About the Renewed US Involvement in Keflavik" (principal investigator: Kousuke Saitou), "Multilateral Political Developments regarding

Security in the Arctic: The Changes in the Views of the US, Russia and Scandinavian Countries and their Effects" (principal investigator: Kousuke Saitou), and "Autonomy and Military Bases in the Arctic: A Relativization of a Viewpoint" (principal investigator: Kousuke Saitou). Thanks to them we were able to conduct several study meetings and collate the texts. In November 2017 at a workshop at the Hokkaido University Institute of Low Temperature Science, Saitou, Kawana, and Koizumi delivered a progress report on this book, receiving huge support from Sumito Matoba, who organized the event, in everything from the composition of the session to the payment of travel expenses. Also, I would like to extend my deep gratitude to principal investigators Shinichiro Tabata and Shin Sugiyama, with whom I worked within the abovementioned ArCS project. The two have helped me immensely, not only with their tough but careful comments but also with funds for travel expenses and equipment purchases.

Incidentally, Kousuke Saitou, who served as the principal investigator for multiple projects, was affiliated with the same laboratory as me when I was a graduate student and is, thus, my direct senior and one of the researchers in whom I place most trust. In fact, the origins of this volume can be traced back to the time we were graduate students in our mid-20s and to the "future research plan" we talked about in a pub near the university campus. This book can be regarded as a progress report on that research.

In writing this volume, we have received the guidance of very many persons, to whom we owe a great debt of gratitude. It is impossible to list them all by name for there are too many. Nonetheless, I am obliged to mention Lill Rastad Bjørst and Ulrik Pram Gad from Aalborg University, where I conducted overseas research from March 2017 to January 2018, who gave me many critical and useful comments from the draft stage of this book (my research at their university was financially supported by the ArCS program for Overseas Visits by Young Researchers). Furthermore, Saitou, Kawana, Koizumi, and I made a joint presentation in December 2017 at the Arctic Politics Research Seminar held in Copenhagen and received beneficial advice from distinguished scholars, such as Hans Mouritzen, Uffe Jakobsen, and Kirsten Thisted, as well as from the abovementioned Ulrik and Lill. In February 2018 my university had the opportunity to host an open seminar, and on that occasion, we had substantial help from the co-organizer the Hokkaido International Exchange and Cooperation Center (HIECC) and their Senior Fellow Yoshihiro Takada. Also, we have received apt comments backed by high scholarly expertise regarding the content of the book and English terms from Ilja Musulin and Adam Worm. I hereby wish to thank them.

I was blessed with the good fortune of meeting Springer's Margaret Deignan during the Arctic Science Summit Week in April 2017 in Prague, Czech Republic, who informed me of the possibility that this book could be published. In addition, during the editing process, we have received tremendous help from Catalina Sava and Truptirekha Das Mahapatra. Our research is still in progress, but we are extremely grateful to Springer for pointing us into the direction of publication.

Finally, on behalf of all the authors, I wish to express our utmost gratitude to our family members for their understanding and support for our research. Wife, husband, child, partner, or parent – their roles are different, but without their presence and understanding for our work, none of us could exist. It is thanks to such support that that we have been able to materialize the present phase of our research in the book before you.

Hokkaido University
Hokkaido, Japan

Minori Takahashi

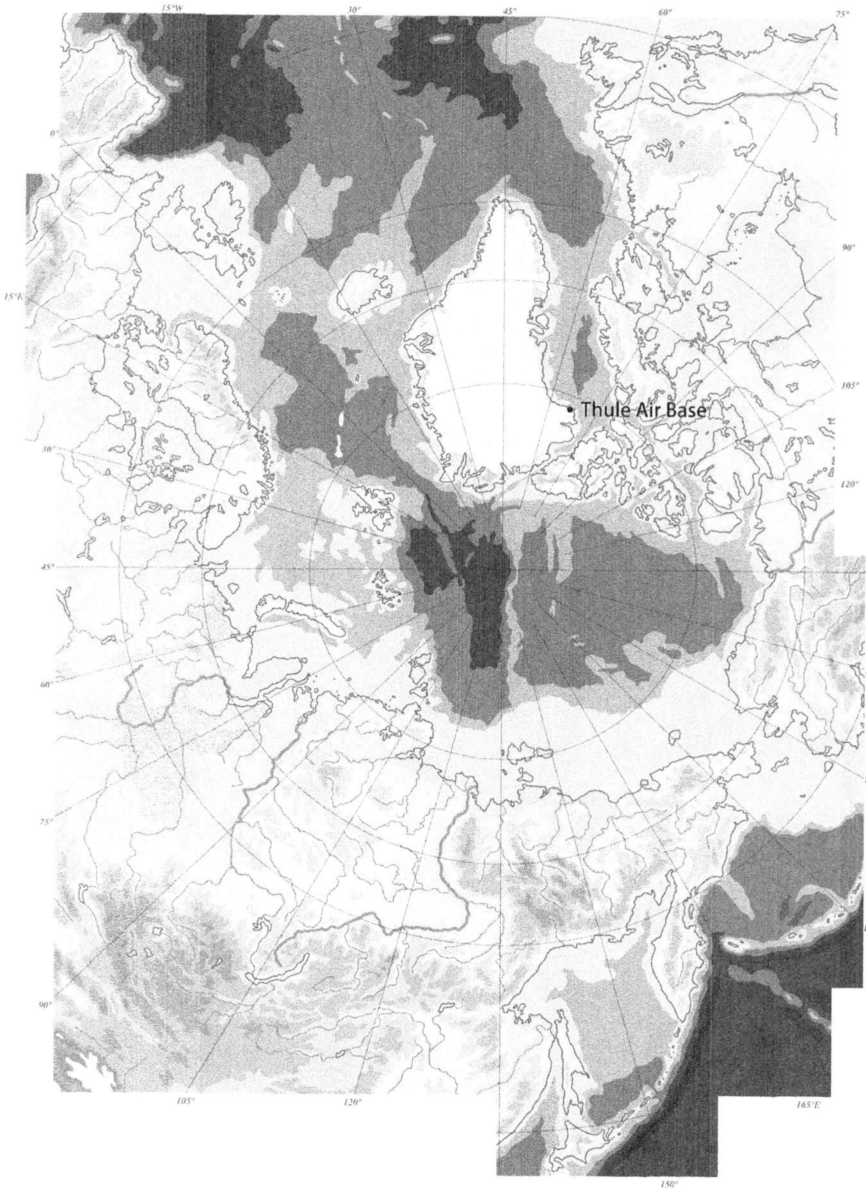
Thule Air Base

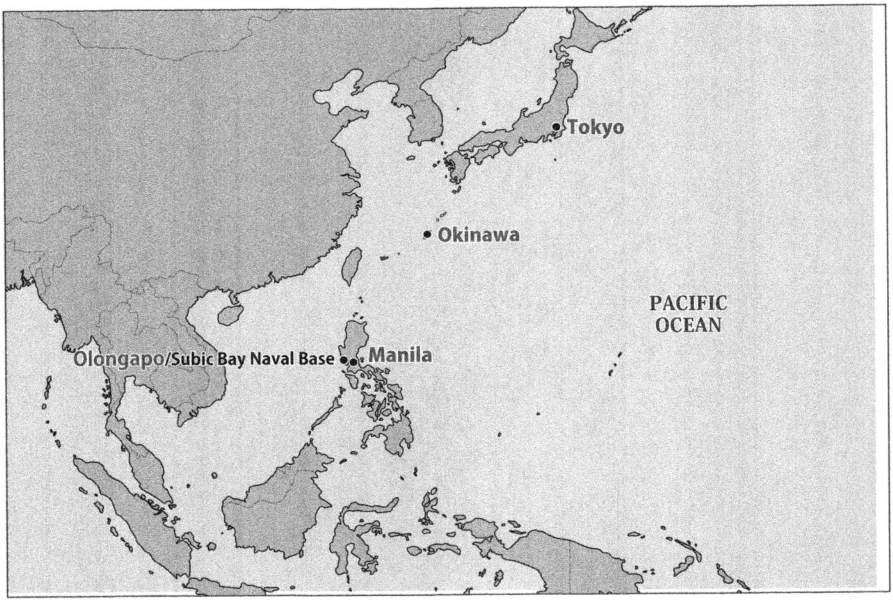
Tokyo
Okinawa
PACIFIC
OCEAN
Olongapo/Subic Bay Naval Base
Manila

Contents

Contributors

Minori Takahashi Ph.D. Assistant Professor, Slavic-Eurasian Research Center and Arctic Research Center, Hokkaido University, Hokkaido, Japan

Shinji Kawana Ph.D. Associate Professor, Institute for Liberal Arts, Tokyo Institute of Technology, Tokyo, Japan

Kousuke Saitou Ph.D. Associate Professor, Research Initiatives and Promotion Organization, Yokohama National University, Kanagawa, Japan

Yu Koizumi M.A. Researcher, The Institute for Future Engineering, Tokyo, Japan

Shino Hateruma, M.A. Ph.D. Candidate, Graduate School of Asia-Pacific Studies, Waseda University, Tokyo, Japan

Ayae Shimizu Ph.D. Researcher, The Institute of International Affairs, Tokyo, Japan

Introduction: The Influence of Sub-state Actors on National Security

Minori Takahashi

Abstract In this volume we shed light on the process in which the sub-state actor of Greenland has strengthened its *de jure* participation in the national security of Denmark. In doing so, we will take up the U.S. Thule Air Base in Greenland as a case that shows that in the relationship between great powers, small countries and local actors within small countries, it is possible for local actors (sub-national entities) to have an influence on higher-level actors in the field of diplomacy on the national security level. For that purpose, we examine political trends involving Greenland, Denmark, the U.S. and Russia by using the multilateral multi-archive approach. We shall also shed light on how the local voice, i.e., the intention to regulate own actions (self-control) and self-rule that concretely embodies it, appear and function in various political matters pertaining to U.S. military bases at the level of national security, by taking up the cases of Okinawa (Japan) and Olongapo/Subic (the Philippines) as reference axes that provide additional insight into the interaction between the U.S. policy regarding overseas basis and the host countries' polities.

Keywords Sub-state actors · National security · Autonomism · Separatism · Greenland/Denmark · U.S. military bases

1 Opening Remarks[1]

Regionalist movements seeking different degrees of autonomy within states or independence have been the subject of numerous debates, with movements for the acquisition of the right to self-rule emerging in West European societies since the 1970s often serving as the case in point (Iwasaki 1992).[2] The goal of the debates has

[1] This part owes a lot to the paper Takahashi 2016: 25–43.

[2] "Autonomy" here refers to the intent to regulate own actions (self-control), as well as to the concrete expression of it (self-rule).

M. Takahashi (✉)
Slavic-Eurasian Research Center and Arctic Research Center, Hokkaido University, Hokkaido, Japan
e-mail: minoritakahashi@slav.hokudai.ac.jp

M. Takahashi (ed.), *The Influence of Sub-state Actors on National Security*, Springer Polar Sciences, https://doi.org/10.1007/978-3-030-01677-7_1

been to shed light on various characteristics of social change that could not be fully grasped using the state-centric framework, through the medium of the region and notions such as autonomy, decentralization and participation. For example, movements demanding greater or lesser degree of autonomy appearing on the periphery[3] of European states, such as those in Northern Ireland, Catalonia, the Basque Country, Bretagne, Occitania, Alsace, Corsica, South Tyrol, Scotland or Wales, have all been taken up many times as study cases for elucidating the character of the relationship between the state and regions (or groups of people within the state), and the power and authority sharing arrangements between them. Such instances of rising regionalisms within the state have been termed "small regionalism", in comparison to the regionalism that transcends individual states, called "grand regionalism". However, "small regionalism" carries within it the constant potential to evolve into "grand regionalism", and activities questioning what the nation state ought to be have, especially after the end of the Cold War, become a trend that is impossible to ignore both in domestic and international politics. For example, the Catalan movement for autonomy and independence from Spain has in recent years made us cognizant of such a cross-over between the small and grand regionalism.

One of the factors behind the manifestation of the desire for autonomy and independence in regions is the creation of uniform policies and the control of the entire demarcated space by the central government (the sole power center), with the goal of maintaining the territorial integrity of the state. Of course, methods for the concentration of power are diverse even under the same form of control, but movements for acquiring the right to autonomy all share a common trait in that that they present a protest against a certain kind of rigidity of the state-centric structure.[4] Generally speaking, central governments (power centers) have, as a way of dealing with such protestation, or in an attempt to cushion it, tended to confer a certain degree of autonomy to sub-state entities (the periphery) and to support decentralization for the purpose of establishing a stable internal, and by extension, international order (Iwasaki 1992: 146–148). For example, law scholar Sun Zhan Kun has focused on

[3] "Periphery" here denotes what has been variously understood as "the invisible" world within the system of the territorial sovereign state. The field of the study of international relations has made extensive attempts to capture the dynamics of the periphery from the following viewpoints. One strain of thought is the criticism of imperialism, i.e., the critique of the dual structure consisting of Western powers (=empires) and non-Western areas (=colonies and subordinated areas). Another is theory of nationalism, which, while taking into account regional differences, criticizes movements for autonomy and independence, which rely on regionalist solidarity, and the postcolonial situation that developed after such movements took place. Third, the Marxist strain, which influenced both the critique of imperialism and the theory of nationalism, and which, based on socialist thought, openly advocates the possibility of socialist revolution coming from the social periphery. Furthermore, today, attempts to understand the periphery from the viewpoint of gender studies, aboriginal studies, or by turning attention to environmental issues are also visible, and this is where the departure point of this book is.

[4] Central governments have tended to award a certain amount of autonomy to sub-state actors to answer or alleviate their protests. The transfers of powers, while similar in structure in the sense that they are conducted between the power center and the periphery, differ in form, as powers can be transmitted within the same organization or from one organization to another, etc.

the fact that international community has given much attention to autonomy as an option for solving or freezing local or ethnic conflicts after the end of the Cold War (Sun 2004). Also, sociologist Takashi Miyajima has offered an analysis, according to which, the rising "need for adapting to the changes in the socioeconomic environment" is one of the factors that has brought about a wave of demands for autonomy and decentralization (Miyajima et al. 2007: 9).

However, the range of inquiry into such demands (the scope of analysis) has often been confined to the category of "internal autonomy", with "external autonomy", which involves the right to conduct diplomacy and thus directly affects state sovereignty, either entirely omitted or insufficiently debated (Takahashi 2013; Ozawa et al. 2016). For example, during discussions about the possible separation of Scotland, as an option in the case that the voters did not support independence, "primary legislative rights in all fields except diplomacy and military affairs", that is, the maximum transfer of powers (devo-max), was placed on the agenda (Kodate 2014). Indeed, if one stands on the premise that the movements for autonomy in West Europe since the 1970s have developed as protests and demands against the rigidity of the political center (Takayanagi 1984: 24), then the reason why diplomatic and military affairs were excluded from the legislative rights was that they were regarded as having the nature of self-determination in external relations and as powers the possession of which is tantamount to separation and independence. In other words, the scope of autonomy has thus far been perceived as confined to internally controlled affairs, and not as encompassing the sphere of interstate relations – the possibility that an internal state constituent could appear in international, diplomatic negotiations as a negotiator (as one of the negotiators) and could influence foreign diplomacy while retaining its own stance has not been sufficiently taken into consideration (Matsui 2011).

If we are to summarize several past studies, we may say that arguments regarding autonomy have ranged between the poles of autonomism and separatism. "Autonomism", most of all, takes into account the economic dependence on the center (the central government), and rather than aspiring for the dissolution of the relationship between the center and periphery, aims to secure survival by adopting positive economic effects and benefits or compromise and strategic cooperation with the center as its priorities. It, thus, chooses to move within the existing state system. Regarding this, Takashi Miyajima and Takamichi Kajita write that autonomist movements, rather than questioning the existence of the state framework itself, aim for "a reexamination of the form of the state by calling for a federation, etc." (Miyajima et al. 1988: 13). Such movements, however, are based on the premise that an asymmetric relationship exists between the center and the periphery and that the periphery will be swallowed by the one-sided, dominant decision making by the state, i.e., they contain a drift to functional localization and require that attention be given both to the intentions and the outcomes.

On the other hand, "separatism" denotes movements that aim to break out of the existing center-periphery relationship and acquire and enhance autonomy in the political field, rather than to make short-term political and economic gains. Separatism emphasizes differences with the center, and involves a break up with the sovereign territorial state and the creation of a new state or community, an ethnic or regional union.

When we look at demands for autonomy at the periphery from the viewpoint of autonomism, we can notice the tendency of such actors to achieve their demands or secure their interests within the sphere of internal relations, while either maintaining "the ruler and the ruled" relationship inherent in the center-periphery relationship or avoiding to further question the character of that relationship. What is characteristic of sub-state actors that act in an autonomist manner is that they show interest in having their position clearly stated in the constitution of the center or the mainland. In effect, the choice of autonomism, although it tends to bring certain gains in autonomy, means settling for a relationship of (at least, legal) superiority and inferiority, as long as the periphery's autonomy is premised on a constitutional guarantee by the center. Despite that, a tendency can be seen in autonomism to opt for maintaining the legal status of a periphery or to avoid a further pursuit of the question of the status, all for the sake of maintaining the living standard and the social order built by the local inhabitants through economic or political dependence on the support of the center.

In contrast, in separatism a tendency exists to call into question the relationship of superiority and inferiority with the center, even if that involves short-term risks (such as a fall in the living standard of the inhabitants),[5] and to aspire for a dissolution of such a relationship and the achievement of self-rule and independence. In separatism, the autonomist option of striving for internal autonomy or the "complete autonomy" within the constitution of the mainland is itself an object of criticism. As a result, the separatist option seeks to relativize and weaken the existing "the ruler and the ruled" relationship with the center.

Thinking in such dichotomous terms tends to highly correlate the expansion of minority[6] rights with the possibility of an internal collapse of the state and to arrive at a zero-sum conclusion. For example, the following remark by the political scientist Gen Kikkawa neatly illustrates how politics has been narrated in the above-mentioned range between autonomism and separatism, premised on the underlying structure in which the periphery is resisting the power center: "Today…the rights of minorities are expanding, and we live in a period in which it is becoming increasingly difficult to maintain the nation state. The expansion of the minority rights means that in the future both the nation state and the territorial state will be constantly exposed to the danger of internal collapse" (Kikkawa 2009: 155).

This way of thinking, which attempts to talk about autonomy by clearly delineating policy areas in which it can or cannot be realized, has been functioning as a thought principle that regulates the narrative concerning diplomacy and military

---

[5] It is worth noting that, as can be seen in the case of Catalonia in Spain, there is a number of instances where the so called "political center" and "the economic center" do not overlap, which is why the movement for autonomy of a political periphery does not necessarily lead to a fall in living standards.

[6] Here "minority" is used as a term that denotes those who are in a weak political or social position and does not necessarily connote literal minority in numbers.

affairs. For instance, the debate regarding the issue of the placement of American military bases in the Danish territory of Greenland, one of the main objects of analysis in this book, has been in many past studies ruled by the logic that the matter was handled solely by the central governments - Washington and Copenhagen. In a number of past studies, the focus was on Copenhagen's thinking, its logic that letting Washington use Greenland (i.e., playing the "Greenland card")[7] secures its own position in the alliance, so the main object of analysis was "a sort of special relationship" between Copenhagen and Washington (DIIS 2005). Of course, there has also been some interest in the internal negotiations between Copenhagen and Nuuk (Greenland) and regional decision-making processes regarding policies (specifically, the decision making in Greenland). However, even then Greenland's role was either explained in terms of it being a part of the Danish state (Kraska 2011), or it was, from the viewpoint of national security, treated as an exceptional case of a region that possesses a special status within the state (Tamnes et al. 2014; Dahl et al. 2014). Or even, the focus was placed on the relativization of the state through autonomy and decentralization (Sørensen 2008). In other words, Greenland's autonomy in the field of national security has been treated as a secondary or a subordinate issue.

As rare exceptions, there have been some attempts to analyze the influence of Greenland (sub-state actors) on international relations by employing the notion of paradiplomacy and identifying the international activities of non-state agents such as local and regional polities (Ackrén 2014). However, in terms of the scope, such research has tended to focus mainly on economic, educational, technological, cultural and environmental cooperation, and on the degree of autonomy that a sub-state actor has in comparison to the central government, and has thus not provided sufficient theoretical and empirical foundations for addressing the sphere of national security, which is important because it can directly affect state sovereignty (Kristensen et al. 2018). For, if, as Urlik Pram Gad (2016: 15) points out "Greenland has taken its paradiplomatic activities to new levels in intimate cooperation with the state formally in possession of its sovereignty", then a need exists to theoretically and empirically elaborate what exactly is meant by the "new levels" and how that "new" dimension has been approached, by taking into consideration not only Greenland, but also the views of mainland Denmark and actors from outside the Arctic area (the U.S. and others).

Building on the above trends and tasks in past research, this volume as a whole sheds light on the process in which the sub-state entity of Greenland has, since the beginning of the debate on its inclusion in the U.S. missile defense shield in the 2000s, strengthened its *de jure* participation in the national security of Denmark. Based on that case, we shall, as different from the past theory of autonomy which has not provided sufficient foundations for the discussion of the commitment of subnational actors to the sphere of national security, demonstrate that the range of

[7] Villaume (1995) points out that the "Greenland card" is a term that enables a realistic and pragmatic understanding of the difficult situation the mainland Denmark is in as a country caught between Eastern and Western powers.

influence of a sub-state actor can exceed the internal matters of a state and encompass international politics, too. Our consideration of that case will, thus, serve to inductively reexamine the thought principle that has been behind autonomy and theory of autonomy thus far.

But, why did Greenland desire *de jure* involvement in national security matters, and how was it able to strengthen its powers in that domain? And, at the same time, why did mainland Denmark hand over to Greenland the powers that might potentially weaken the cohesiveness of the Danish state? What was the strategic concern of the United States, the country that established the military base in Thule, and what was its view of Greenland, or rather, security in the Arctic? What was the reaction of Russia, as a regional power, to such developments? How did Greenland, which hosts the base, understand its relationship with the U.S. and Denmark? In order to shed light on these questions, in this volume we will take up the U.S. Thule Air Base in Greenland as a case that shows that in the relationship between great powers, small countries and local actors within those small countries, it is possible for local actors (sub-national entities) to have an influence on higher-level actors, and will endeavor to elucidate trends concerning Greenland, Denmark, the U.S. and Russia by using the multilateral multi-archive approach. In doing so, we shall shed light on how the local voice, i.e., the intention to regulate own actions (self-control) and self-rule that concretely embodies it, appear and function in various political matters pertaining to U.S. military bases at the level of national security, by taking up the cases of Okinawa (Japan) and Olongapo/Subic (the Philippines) as reference axes that allow additional insight into the interaction between the U.S. policy regarding overseas basis and the host countries' polities. To further elaborate on the role of the chapters on Okinawa and Olongapo in this volume or the reason why their cases have been chosen we can add the following: in order to examine contrasting political choices in relation to Greenland, we take up Okinawa, which has missed an opportunity to have its demands accepted, as an example of a failure of a sub-state actor to exercise influence on national security, and Olongapo, where the voice of the locals functioned to constrain the actions of the central government resulting in the withdrawal (or eviction) of the U.S. military, as an example of success. If we assume that a relationship of superordinacy and subordinacy exists between the chapters, then the chapters on Okinawa and Olongapo serve as a supplementation or reference axes for elucidating the case of Greenland, and thus cannot be central. Nonetheless, it goes without saying that these chapters are indispensable in the sense that they play an important role in dispelling the seemingly naïve but essential doubt that the readers of this volume might have: why is the topic of this book autonomy and military bases (national security), and why is it that Asians, that is, Japanese scholars are thinking about that issue?

2 Composition of the Book

The work process for achieving the above goals will unfold as follows. We first define the scope of this book with a view to setting a theoretical framework for discussing the U.S. policy concerning overseas military bases, as well as the responses of countries that have accepted them and the subnational entities in which the bases are located (Chapter "Base Politics and the Hold-Up Problem" by Shinji Kawana). We then trace the history of Greenland from World War II to Cold War, where a U.S. base has been installed and operated, and empirically demonstrate how Greenland expanded its autonomy through the case of the incorporation of that base, the Thule Air Base, into the U.S. missile defense shield in 2000s by looking at the debates in Greenland's and Denmark's parliaments (Chapter "Greenland's Quest for Autonomy and the Political Dynamics Surrounding the Thule Air Base" by Minori Takahashi). We also reexamine the influence of sub-state actors on diplomacy using the framework of the theory of autonomy. Furthermore, we also shed light on the dynamics in which Greenland, despite facing the reality of not being necessarily able to exercise its autonomy rights sufficiently, is, together with the convulsions in "Arctic politics" that stem from recent climate changes, the delineation and demarcation of borders and the utilization of natural resources and new sea lanes, turning into an actor that cannot be ignored not only in domestic but also in international politics - the dynamics in which the purpose of the Thule base is diversifying and can now function as a card for negotiations regarding Greenland's autonomy.

Building on that argument, the volume then proceeds to include as explanatory variables the evolving interests of the U.S., which operates the Thule military base, together with the policy making process in the country (and area) hosting the base and the political dynamics of mainland Denmark. Also included is the Russian strategy concerning military bases and the Arctic, since that country has been intensifying its military activities by deploying and stationing submarines, building naval basis and refurbishing the existing military installations in the Arctic area (Chapter "How Have the U.S. Interests in Greenland Changed? Reconstructing the Perceived Value of Thule Air Base after the Cold War" by Kousuke Saitou, and Chapter "Russia's Military Build-up in the Arctic: Russia's Threat Perception and its Military Strategy in the Arctic Region" by Yu Koizumi). This is because the presence of the Thule Air Base cannot be explained by looking only at the decision-making process within Greenland, as it is also influenced to a large extent by the political dynamics in the entire Denmark and by international political trends regarding military bases.

Furthermore, by examining the cases of Okinawa and Olongapo, which serve as points of reference for our analysis of the interaction between American policies regarding military bases and the political systems of the host countries, we shed light on how the local voice appears and functions in a variety of political issues concerning military bases, that is, how it functions in the sphere of national security (Chapter "Okinawa's Search for Autonomy and Tokyo's Commitment to the Japan-U.S. Alliance" by Shino Hateruma, and Chapter "The Political Dynamics and Impacts Surrounding Subic Naval Base in the Philippines" by Ayae Shimizu). We believe that such efforts in this volume will ultimately lead to clarifying the dynamism of international security, as seen from the viewpoint of internal constituents ("Conclusion: The political choices of substate actors and the politics surrounding U.S. military bases" by Minori Takahashi).

In this volume, to empirically elucidate the process of the intensification of the *de jure* involvement of sub-state actors in the field of security, we shall historically examine past diplomatic material, utilize parliamentary minutes for synchronic analysis, and with that combine interviews with policy makers and military base personnel. More than political bargaining between states, what we aim to clarify, by looking at the field of national security and by focusing on diplomatic and military matters, that have traditionally been the exclusive jurisdiction of the state, are policy choices of sub-state actors. In other words, in past studies there has been a stark tendency to approach internal constituents of states as entities who simply question to what extent agreements reached through international negotiations should or should not be accepted, but our research will reveal the existence of an internal constituent that is getting actively involved in international negotiations as one of the negotiators (together with the central government), while continuously maintaining its own position.

This volume starts from an inductive perspective, but, through the theoretical framework that includes in its scope the interrelation between U.S. politics regarding overseas military bases and autonomy, provides the possibility of joining that perspective with deductive research (Chapter "Base Politics and the Hold-up Problem"). It also advocates the necessity of interaction between inductive and deductive approaches for making visible the factors behind the fluctuating security situation in the Arctic, where a latent potential exists for the confrontation of various, mutually opposed interests. Issues concerning political stability in the Arctic region are closely related to the traditional security strategies and political dynamism that are symbolized by the strategies of great powers regarding military bases, which this volume seeks to clarify. The present book aims to make an academic and social contribution by providing a new perspective and material for analyzing security issues in the Arctic.

References

Ackrén, M. 2014. Greenlandic paradiplomatic relations. In *Security and sovereignty in the North Atlantic*, 42–61. Basingstoke: Palgrave Macmillan.

Dahl, A. et al. 2014. *Northern security and global politics: Nordic-Baltic strategic influence in a post-unipolar world*. London/New York: Routledge.

Danish Institute for International Studies. 2005. *Danmark under den Kolde Krig*.

Gad, U.P. 2016. *National identity politics and postcolonial sovereignty games: Greenland, Denmark, and the European Union (Monographs on Greenland)*. Copenhagen: Museum Tusculanum Press.

Iwasaki, M. 1992. Renpou-shugi to renpou-sei [Federalism and the Federal System]. *Administrative Studies Bulletin* 27: 141–166.

Kikkawa, G. 2009. Kokusai mondai toshiteno mainoriti [Minorities as an International Issue]. In *Japan's international political science Vol. 2-international politics without borders*. Tokyo: Yuhikaku.

Kodate, N. 2014. "Bunri dokuritsu" wo tou sukottorando juumin touhyou: "Kurashimuki" ka "aidentiti" ka [The Scottish Referendum on "Independence": "Good Living Standard" or "Identity"?]. Synodos. URL=https://synodos.jp/international/7439.

Kraska, J. 2011. *Arctic security in an age of climate change*. New York: Cambridge University Press.

Kristensen, K.S. et al. 2018. *Greenland and the international politics of a changing arctic: Postcolonial paradiplomacy between high and low politics*. London/York: Routledge.

Matsui, Y. 2011. *Kokusaihou kara sekai wo miru: Shimin no tame no kokusaihounyuumon [Looking at the world from the perspective of the international law: An introduction to the international law for citizens]*. Tokyo: Toshindo.

Miyajima, T. et al. 1988. *Gendai yoroppa no chiiki to kokka: Henyou suru cyuushin-syuuhen mondai heno shikaku [Regions and states in contemporary Europe: A perspective on the changing issue of the center-periphery relationship]*. Tokyo: Yushindo Kobunsha.

———. 2007. *Chiiki no yoroppa: Tasouka, saihen, saisei [The Europe of regions: Stratification, realignment and revival]*. Tokyo: Jinbun-Shoin.

Ozawa, M. et al. 2016. *Aisurando, guriinranndo, hokkyoku wo shirutameno 65 shou [Sixty-five chapters for knowing Iceland, Greenland and the Arctic]*. Tokyo: Akashi Shoten.

Sun, Z.K. 2004. Kokusaihou ni okeru "jichi" no gainen to sono kinou [The notion and function of "autonomy" in the international law]. *Nagoya University Journal of Law and Politics* 202: 43–78.

Sørensen, A. 2008. *Denmark-Greenland in the twentieth century (Meddelelser om Grønland)*. Copenhagen: Museum Tusculanum Press.

Takahashi, M. 2013. *Jiko ketteiken wo meguru seijigaku: Denmaaku-ryou guriinrando niokeru "taigaiteki jichi" [Politics over self-determination rights for the external self-governance of Greenland, a territory of Denmark]*. Tokyo: Akashi Shoten.

Takahashi, M. 2016. The politics of the right to self-determination: Reframing the debate on Greenland's autonomy. *Eurasia Border Review* 6 (1): 25–43.

Takenaka, C. 2007. Shuen kara no kokusai seiji [International politics as seen from the periphery]. *International Politics* 149: 2–3.

Tamnes, R. et al. 2014. *Geopolitics and security in the Arctic: Regional dynamics in a global world*. London/New York: Routledge.

Villaume, P. 1995. Thule og den danske selvforståelse. *Udenrigs* 50 (3): 29–36.

Base Politics and the Hold-up Problem

Shinji Kawana

Abstract In this chapter, we will consider the conditions under which local actors exercise political influence on their governments as part of base politics during peacetime in democratic states. A factor that destabilizes bases in democratic states is the "vulnerability" of host governments. Vulnerability is as a function of substitutability, urgency, and specificity. The greater the importance of a particular base for the government, the greater its vulnerability *vis-à-vis* local political actors, and the higher the risk that it will be "held up" by them. However, as the government knows the incentives for opportunism among these local actors, it will aim to integrate, institutionalize, and distribute the base transactions in order to avoid a hold-up.

Keywords Vulnerability · Substitutability · Urgency · Specificity · Integration · Institutionalization · Distribution

1 Introduction

The United States' overseas bases have always faced operational difficulties during peacetime.[1] It is often pointed out that a base's political stability tends to be strong when the political system of the receiving country (hereinafter, the host state) is authoritarian, and there have been times when the United States' policy of global democratization has seriously strained relations (Calder 2007; Stein 1963). After the Cold War, the United States embarked on full-fledged efforts to tackle this

[1] During the Cold War, bases became politically unstable in Turkey, Greece, the Philippines, Portugal, France, and Ireland. After the 2000s, political problems frequently occurred at bases such as those in Turkey, South Korea, and Japan (Okinawa). Kawato, Y. 2015. *Protests Against U.S. Military Base Policy in Asia: Persuasion and Its Limits*, Stanford, California: Stanford University Press; Vine, D. 2015. *Base Nation: How U.S. Military Bases Abroad Harm America and the World*, New York: Metropolitan Books, Henry Holt and Company.

S. Kawana (✉)
Institute for Liberal Arts, Tokyo Institute of Technology, Tokyo, Japan
e-mail: shinji.kawana@ila.titech.ac.jp

M. Takahashi (ed.), *The Influence of Sub-state Actors on National Security*, Springer Polar Sciences, https://doi.org/10.1007/978-3-030-01677-7_2

problem, and one of the main themes in its Global Posture Review (GPR) was the stable management of its bases in democratic states.[2] The so-called "rebalancing" policy of the Obama administration emphasized securing the political stability of bases in the Asia-Pacific region, one of the premises of the policy as a whole.[3]

On the other hand, it is not easy for host states to continually reinforce the political legitimacy of the United States maintaining bases on their soil. To achieve this, there are times when they must significantly change their allocation of political resources (e.g., subsidies or increased autonomy), and they must at times "leave out" actors from the administration. Particularly when the system of the host state is democratic, it is often the case that specific local government actors (i.e., local councils) have *de facto* "veto power" over the operations of bases. Thus, such governments are frequently faced with problems concerning domestic base politics.

This problem of the "compatibility of democracy and military bases" constitutes one of the debates in contemporary international relations theory. For example, Alexander Cooley argues that if the host state is a mature democratic nation, the "base contract" is highly reliable, and if that country relies on the United States for its security, the base will be politically stable (Cooley 2008). Conversely, Kent E. Calder argues that in a democratic nation, in every aspect of the multilayered decision-making process, bases are easily destabilized because they are affected by domestic political factors (for example, public opposition to the bases, base-related incidents and accidents, population density, and norms).[4]

In this way, the two studies present contrasting hypotheses about the effects of the host state's political system on a base's political stability, but the debate is not limited to this point. For example, according to Cooley's framework, countries such as Japan and South Korea, both of which are democratic and equally dependent on the United States for security, may be indifferent to bases' stability. Alternatively, at the government level, the base contract will be considered stable (e.g., this seems to apply to Japan–U.S. relations since the 1990s) as long as it is fulfilled no matter how politically problematic the base becomes domestically. In addition, in terms of the domestic factors that Calder extracted as independent variables, it is not certain when and under what conditions they will have a negative effect on a base's stability, even if the validity of each one is intuitively acknowledged. Alternatively, even if these factors have a decisive influence on certain aspects of domestic politics, it is not theoretically self-evident what the relationship between them and the domestic political system would be.

---

[2] Statement of Douglas J. Feith, Under Secretary of Defense for Policy before the House Armed Services Committee, June 23, 2004: 4–5.

[3] At that time, Secretary of State Hillary Clinton commented on the new national defense strategy (i.e., the Asia shift), and in this context, the question of whether bases were politically sustainable was raised. Hillary Clinton, "America's Pacific Century: The Future of Politics Will be Decided in Asia, Not Afghanistan or Iraq, and the United States Will be Right at the Center of the Action," *Foreign Policy,* November 2011: 1–14.

[4] Calder, *op.cit.*

On this point, Andrew Yeo used the concept of "security consensus" and seems to have overcome Calder's problem (Yeo 2011). According to Yeo, whether domestic opposition to a base and base-related incidents and accidents turns into government-level political problems will depend on the strength of the security consensus in the host state. This is measured by (1) the depth of the policy elite that attempts to maintain a security relationship with the United States and (2) the extent to which the maintenance of that relationship is embedded within the existing social system. In other words, if a broad domestic policy elite maintains a positive attitude toward the security relationship with the United States, and if the security relationship with the United States has been fully institutionalized, then the political stability of the bases will be strong. However, when this framework is used, for example, within a specific host state, it does not explain how some bases are politicized and others are not, or the differences between those local governments that are able to influence the central government and those that are not.

Therefore, in this chapter, we will consider the conditions under which local actors exercise political influence on their governments as part of base politics during peacetime in democratic states. I hope thereby to identify the factors that determine the differences that previous studies did not explain, namely, the differences between cases of stably operated bases and cases of unstable bases in democratic countries as well as the differences between local political actors that influence the central government and those that do not.

It has been found that one factor that destabilizes bases in democratic states is the "vulnerability" of host governments, which is inherent to base politics. Vulnerability is measured in terms of the characteristics of any base that is "transacted" between the government and local actors; in other words, it is expressed as a function of substitutability, urgency, and specificity. Put differently, the greater the importance of a particular base for the government, the greater its vulnerability *vis-à-vis* local political actors, and the higher the risk that it will be "held up" by them. However, as the government can also predict the incentives for opportunism among these local actors, it will aim to "integrate," "institutionalize," and "distribute" the base transactions in order to avoid a hold-up.

The local actors that are addressed by the model in this chapter are rational actors trying to improve their utility (e.g., their political and economic interests) by accepting the presence of bases. Therefore, the scope of the model's application is limited to base politics that involve the democratic governments of sending and host states that have an interest in maintaining a base, as well as local political actors that opportunistically choose their political stance with regard to that base (e.g., whether to allow it or to refuse it). As such, this model cannot explain the decision-making process and behavior of so-called "non-opportunistic" local actors that may refuse bases *a priori*.

In Sect. 1, we will examine the relationships between the three parties that influence base politics in democratic states: the sending country, the host government, and the local political actors. We will also consider the "vulnerability" and "hold-up problem" that have an effect on these relationships. In Sect. 2, the vulnerability of the government is identified as a function of the substitutability, urgency, and specificity, which are transactable base assets. In Sect. 3, we will consider potential government countermeasures for this vulnerability, or put differently, the problem of vulnerability control.

2 Base Politics

2.1 A Triadic Relationship

In democratic states, base "transactions" involve the central government and local political actors, meaning that they are not simply transactions between two parties (a dyad). It goes without saying that the host government generally supplies the sending state with a base with the approval of local political actors, on the basis of an understanding (treaty or agreement) reached by the governments of the sending and host states. In other words, at the very least, three political actors are involved, and the amounts of political resources owned by these constitute the dynamics of the base politics.[5]

On this point, as noted by Georg Simmel, some aspects of a triadic relationship cannot be reduced to a dyadic one (Simmel 2016). For example, as soon as a dyadic relationship is expanded to a triadic relationship, the power of the subordinate party in the earlier dyad is redefined by the other two parties, that is, by the terms of the relationship between the intermediate and the superior parties. Now, if we have a local actor and a central government (a dyad) negotiating a base, what would happen to that relationship if it is expanded to a triadic one that also includes the sending states? (From the viewpoint of international relations theory, this crosses the boundary between the international and the domestic level, but does not affect the essence of my analysis.) We can assume asymmetry in the powers of the local political actors and the government, and that there is asymmetry in the powers of the government and the sending state. In other words, we assume that the sending state is in a superior position in relation to the host state's government (e.g., imagine the relationship between the United States and any host state after World War II). As a result, the relationship between two parties, normally the relationship between local actors and the government, which should be defined as superior and subordinate (or in some cases, "dependent" and "controlling" (Johnson 2004)), is rearranged into a triadic relationship of subordinate (local political actors), intermediate (host government), and superior (sending state).

In this triadic relationship, whether the relationship between the superior and the subordinate actors is direct affects the power of the subordinate. According to

[5]A paper that incorporates this sort of triadic relationship into an analysis of base politics is, for example, Archer, C. 2003. "Greenland, US Bases and Missile Defense: New Two-Level Negotiations?" *Cooperation and Conflict*, Vol.38, No. 2: 125–147. For an analysis of international relations in general, see Putnam, R. 1988. "Diplomacy and Domestic Politics: The Logic of Two-Level Games," *International Organization*, Vol.42, No. 3: 427–460.

Simmel, when a relationship between a superior and a subordinate is open, the intermediate is in a more disadvantageous position than when it is closed, and this improves the subordinate's position. In other words, the intermediate, who is inserted between the superior and the subordinate, is obviously superior to the subordinate, but what is important here is whether the subordinate party can become the superior. In this sort of relationship, the relative strength of each party is more flexible, while in a dyadic relationship it tends to be fixed.[6]

This structure can be seen in the triadic relationships between the United States, the governments of host states, and local political actors as they conduct base transactions. In other words, what determines the relative strengths of the three political actors is whether the local political actor can form a direct relationship with the actor who has the superior position among the three (that is, the United States). The conversion from dyadic base politics, in which power relationships tend to be fixed, to highly fluid triadic base politics can increase the influence of local political actors on the government.

To repeat, in the base politics of democratic states, for a base to be kept open, the government needs the approval of the local political actors (or residents), both procedurally and substantively. On the other hand, even if the local actors want the base to be shut down, this is not easily realized with just their own political resources. For example, in countries in which the government has secured a strong consensus on the security relationship with the superior actor (i.e., the United States), the anti-base movement is less likely to penetrate the host government's decision-making.[7] However, even in cases in which they do not have direct influence (or do not actually participate in the base transaction), they are able to send negative signals to the superior by impeding the stable operation of the base or various acts of political omission.

For example, Japan's local governments are sometimes understood to have a "veto" when it comes to the more important aspects of base politics. To give one example, supposing that the Japanese government decided that it wanted to reclaim an area of the sea to construct new bases, the local governor can effectively use the Act on Reclamation of Publicly Owned Water Surface to "veto" this. If the governor exercises such authority, or even only hints at the possibility of doing so, the subsequent base politics will be transformed from a dyad to a triad. For many host-state governments, providing a base is an obligation stipulated in a treaty or agreement with the sending state. Failure to fulfill it will strain the security relationship with the sending state. Straining the relationship with the United States is likely to have a certain influence on the host-state government, which finds value in that

---

[6] In this sense, there is a certain logic to the fact that the subject of the analyses in research on base politics that adopt dependency theory as a paradigm is mainly dyadic relationships. See, for example, Yamazaki, T. 2010. *Politics, Spaces, and Places: Towards a Geography of Politics,* Kyoto: Nakanishiya Shuppan (In Japanese). For a more general discussion, please refer to Rokkan, S. and D. W. Urwin. 1983. *Economy, Territory, Identity: Politics of West European Peripheries,* London: Sage Publications.

[7] Yeo *op.cit.*: 27–32.

security relationship. Keeping watch on the reactions of the sending state, the government is required to handle its relationship with the subordinate more carefully than ever in order not to worsen the "mood" of its superior. In other words, the government must show consideration to local actors outside of domestic political concerns, on the basis of international political concerns. Alternatively, if the sending state also reacts sensitively to pressure from local actors and wants to avoid their opportunism, like the host-state government, it will probably use various political resources to try to influence the intentions of local actors (e.g., returning minor bases or reviewing SOFA). In other words, whether it wants it or not, the superior party becomes involved in the "blackmail" of the subordinate parties in base politics, and in some cases, this has the effect of "bypassing" the influence of the intermediate party.

2.2 Vulnerability and the Hold–Up Problem

The concept of "vulnerability" is useful for understanding this sort of relationship between the government and local actors (Goldmann and Sjostedt 1979). In the context of international relations theory, vulnerability is generally defined as "the cost that is paid to effectively adapt to a changed environment" (Keohane and Nye 2001). For example, after A acts, the higher the cost that B pays to adapt to the new environment that follows from that action, the more vulnerable B becomes in relation to A. When the amount of vulnerability differs relatively, an asymmetric relationship is created between the two parties. In other words, the side with the higher vulnerability (not the government or the region) is placed in the weaker position.[8]

As noted by Albert O. Hirschman, if a clear power relationship exists between a controlling side and a subordinate side, where the former obtains (exploits) enormous interests from the latter, this also means that the beneficiary (controller) is already dependent on the provider (subordinate).[9] That is to say, for the side that receives relatively little benefit from the transaction, threatening to interrupt the transaction is a weapon against the other, superior party. With this weapon, the weaker party can acquire superiority in terms of power (independent of the amount of its material power) by increasing the extent of the stronger party's reliance on itself.

[8] However, some argue that the degree of dependence does not affect the influence of the two parties. For example, see Wagner, R. H. 1988. "Economic Interdependence, Bargaining Power, and Political Influence," *International Organization*, Vol.42, No. 3: 461–483.

[9] Hirschman, A. O. 1980. *National Power and the Structure of Foreign Trade*, expanded edition, Berkeley: University of California Press. For a paper developing a similar argument, see Knorr, K. and F. Trager, eds. 1977. *Economic Issues and National Security*, Lawrence, Kansas: University Press of Kansas.

This phenomenon of the reversal of a power relationship that occurs based on an imperfect transaction relationship is known as the "hold-up problem."[10] It is fully possible that this phenomenon occurs between a government and local political actors conducting a base transaction.[11] Let us assume that at the present time, the United States provides a certain state with security in return for that country providing bases to the U.S. (in fact, many of the countries that the United States established as allies after World War II had such an agreement (Sandars 2000)). Now, if the relevant host state cannot guarantee its own security without the United States, the relevant government relies not only on the United States, but also on the local governments that accept the domestic bases. At this time, the relevant local government may "hold up" various aspects of the government's domestic base politics. Under certain conditions, the local actors have room to engage in opportunistic behavior toward the government. For example, local actors may demand that the government changes the base's transaction conditions (e.g., increased subsidies, noise-reduction measures, reduction of the base size, expansion of the authority of local political actors, etc.), and if these demands are not met, it can threaten to take actions that will damage the stability of the base (e.g., not approving land use, and the local government's assertion of political independence from the central government). In such a case, the government will be forced to judge whether it should meet their demands or refuse them based on the value of what has been taken "hostage."

3 Substitutability, Urgency, and Specificity

What then determines the effect (the success or failure) of the hold-up by local actors? Conversely, what elements constitute the size of the central government's vulnerability? Below, we shall consider this question from the base characteristics that are transacted between the two parties or, more specifically, by breaking them down into substitutability, urgency, and specificity.[12]

---

[10] Studies applying the hold-up problem to political science include, for example, Carnegie, A. 2014. "States Held Hostage: Political Hold-up Problems and the Effects of International Institutions," *American Political Science Review*, Vol.108, No. 1: 54–70; Lake, D. A. 1999. *Entangling Relations: American Foreign Policy in Its Century*, Princeton, New Jersey: Princeton University Press; Wallander, C. A. 2000. "Institutional Assets and Adaptability: NATO After the Cold War," *International Organization*, Vol.54, No.4: 705–735.

[11] For more on the politics of overseas bases, with regards to the power-reversal phenomenon between the sending states and host state, or in other words, the relationship at the governmental level, refer to Liska, G. 1962. *Nations in Alliance: The Limits of Independence,* Baltimore: The Johns Hopkins Press: 116–160; Cooley, A. and H. Spruyt 2009. *Contracting States: Sovereign Transfers in International Relations*, Princeton, NJ: Princeton University Press; and Kawana, S. 2012. *Bases Politics – The Origins of the United States' Post-War Overseas Bases Expansion Policy,* Tokyo: Hakuto-Shobo (In Japanese).

[12] For more on this topic, please refer to Kawana, S. 2014. "Bilateralism and its Application to Vulnerability," *The Law Review of Kinki University,* Vol.61, No. 4: 1–26.

3.1 Substitutability

Substitutability refers to the possibility of the government procuring domestically a base of the same value in the event that a certain base transaction is interrupted. For example, if we imagine a case in which the government transacts with local government A for an airfield and with local government B for barracks, and both are interrupted at the same time, the damage from the transaction with A will be larger. This is because the potential tolerance values for each local government will differ depending on the purpose for which the land and facilities are used. Of course, the base's geographical attributes will also matter. For example, the geographical proximity of countries considered to be a threat will affect the substitutability of an airfield. In other words, after the government has suspended transactions with A for an airfield in a strategically important location, it will not be easy for it to procure an airfield of equal value through a transaction with another local actor. Thus, in terms of substitutability, the government's vulnerability can be said to be higher in relation to A than B (Knudsen 1979).

3.2 Urgency

A base with a high level of urgency is a base provided by the host government that is recognized as being indispensable for the survival and prosperity of the host state and that is also highly evaluated for its strategic importance by the sending state. For example, accepting the base of a world power (an allied state) during wartime is highly urgent for the host state. Therefore, the urgency of this type of base is normally higher during wartime than peacetime. However, during emergencies, the security awareness of the government and the local political actors normally align, so it is difficult to believe that this vulnerability problem would have a serious impact on the government during such a time.

On the other hand, even during peacetime, which is the main subject of the analysis in this paper, if both the host state's government and the sending state evaluate the base as strategically important and consider its maintenance as foundational to their alliance, the urgency of that base will be high. For example, airfields, communication sites, missile and radar facilities, and other such facilities near a country that is a potential threat have a deterrence effect or serve to symbolize an alliance. Considering this, urgency would remain high even during peacetime. On the other hand, facilities such as barracks or logistical facilities used to deploy ground troops are not necessarily so urgent.

3.3 *Specificity*

In micro-economics, specificity refers to "value that becomes extremely high only in certain specific relationships and circumstances," (Milgrom and Roberts 2003) and such goods are called "specific assets."[13] Here, as a transaction good, a base is an asset that creates value from having a specific purpose or from its sole use for military and security purposes. Moreover, in practical terms, it is difficult for the investing side (for the purposes of this discussion, this refers to the host state's government only) to change this purpose to a different one (e.g., for civilian use).[14]

In addition, the "difficulty of transfer" of an asset that is physically enormous is also a factor that determines its specificity (Teece 1980). For example, supposing that the sending state returned the training fields and helipads located in the mountains, it would be difficult to use them as civilian facilities or to redevelop the land on which they were constructed. In the same way, it is not realistic to transfer airfields that are located in deserts or on ice fields to civilian use. Thus, naturally, it must be said that their level of specificity is high. By contrast, it is possible to use airfields or port facilities within or near urban areas as civilian airports or port facilities, or for the land to be redeveloped, after they have been vacated by the military of the sending country.

If a base has a high level of specificity, this reverses the strength of the parties involved in the transaction (other conditions being equal). This is because the government tends to excessively prioritize the stable operations of an already-constructed base. Thus, if it is faced with opportunistic behavior from local actors (e.g., demands to review the SOFA or to increase land-usage fees) after the transaction has been concluded, it has no choice but to accept their demands. On the other hand, when the base does not have specificity and it is possible to convert it to other uses, the government is able to refuse the demands of local actors. David A. Lake notes that "[t]he more relationally specific the asset, the greater are the quasi rents and, in turn, the costs to the polity if a partner should act opportunistically."[15] Wallander also concluded that transactions of goods with a high level of specificity are affected by opportunity costs when economic and political actors adopt a substitute strategy.[16]

We have examined three factors above that, on the one hand, constitute the vulnerability of the government, and on the other hand, affect the success or failure of local actors' hold-ups. However, these three factors are interrelated, and even if it is assumed that one factor is a given, the extent of the vulnerability can vary considerably depending on variations in the other factors (in particular, we should be aware

---

[13] Wallander, op.cit.

[14] The extent of specificity is measured by the rate of decline in value that occurs when using the asset for purposes other than the original purpose. For the relationship between bases and specific assets, refer to Lake, D. A. 2001. "Beyond Anarchy: The Importance of Security Institutions," *International Security*, Vol.26, No. 1: 129–160.

[15] Lake, *op.cit.*: 54.

[16] Wallander, op.cit.: 708.

that the level of urgency is highly dependent on the strategic environment of the sending and host states).

4 Integration, Institutionalization, and Distribution

Under certain conditions, there are certain incentives for opportunistic behavior by local actors in base politics, and the government also understands this. Therefore, it would seem that the government, in order to control its own vulnerability (i.e., to mitigate hold-up risks), will attempt to appropriately design the base transactions with local actors. If we take the perspective of micro-economics, the side attempting to design the transaction may try to procure the goods internally, rather than from the market (Carroll and Teece 1999). This is called "integration." In some cases, as an intermediate form between market and integration, it may attempt to lower the incentives for opportunism by concluding an institutional transaction (a so-called long-term contract or relational contract). Alternatively, by distributing the parties involved in the transaction and repeating a spot contract (a contract that does not require any specific investments, that is concluded every time the base is used, and for which usage fees are renegotiated on every occasion), the government might try to reduce the cost of the opportunism per transaction until it is low enough to be ignored. How can these views be applied to an analysis of domestic base politics?

4.1 Integration

First, the government may be able to reduce its vulnerability by depriving the other party of "residual rights of control" (Williamson 1975; Cooley 2005) If a dyadic relationship is horizontal, then the residual rights of control will belong equally to all parties. However, in a more hierarchical relationship, the more one side has residual rights of control, the more the other side will lack them. According to Lake, in a horizontal relationship between two parties, if both parties have complete residual rights of control, each will possess sufficient discretion for their future behavior, or in other words, for opportunism (a policy option). On the other hand, the smaller their rights, the less room they will have to engage in opportunistic behavior.[17]

Therefore, integrating a transaction is an effective method for avoiding the other party's opportunistic behavior. That is to say, a government can reduce its vulnerability by integrating a colony and incorporating it as a directly controlled administrative district (prefecture). This is to invalidate the opportunistic behavior of the other party through ownership of the good. Some examples that seem to correspond to the "integration" of base transactions described here include Greenland by

[17] Lake, *op.cit.*: 53–57.

Denmark from 1953 to 1978, the Azores by Portugal from 1976 onwards, and Diego Garcia by the United Kingdom from 1966 onwards (in each case, a base was established that was important for both the United States and the host state's government) (Vine 2009).

4.2 Institutionalization

"Institutionalization" would seem to be a second method for governments to control their vulnerability.[18] For this, the government tries to build a continuous relationship based on mutual interests between itself and local actors. In terms of policy, Calder calls it "compensation politics,"[19] which often appear in the form of grants and subsidies, side payments, and positive sanctions *vis-à-vis* local actors (Friman 1993). An effective combination of these methods by the government can enable it to switch from being unilaterally "held hostage" to a situation in which both parties are "mutually held hostage" by each other (i.e., the "symmetrization of vulnerability") (Yamamoto 1989).

Viewed from the perspective of the local political actors, even supposing that they hold up the government in base politics and their initial conditions are temporarily improved, they face the risk of government retaliation in other areas. Moreover, in the event that the "system" is explicitly made into law (legalization), any violation of the law may adversely affect their reputation and their trustworthiness within society. Under a repeating game, it is possible that such a violation may backfire and lead to larger costs in the long term. As a result, it would reduce the incentives for local opportunism.

For example, Japan has budgeted more than 400 billion yen annually from the 2000s for the costs of institutionalizing base transactions. Broken down, these figures include soundproofing work, such as in the residential areas surrounding bases; "expenses for vicinity improvements" to supplement the spending on public facilities by local governments; and "facilities' rental expenses, compensation, etc.," as the fees to rent the land or to compensate fishermen for the restriction or prohibition of operations by fishing boats. They also include expenses relating to SACO (Special Action Committee on Okinawa) at the government's discretion for the selection of bases and local governments for decisions on the provision of subsidies, base subsidies, and base compensation (Kawase 2013). These were all side payments linked to base transactions.

Moreover, this system was legally supported by the Ministry of Defense (formerly the Defense Facilities Administration Agency), which acted as a mediator with local actors. The Ministry of Defense has regional bureaus that deal with complaints by citizens, negotiate the rental expenses for the land provided for the base,

---

[18] Carnegie, op.cit.

[19] Calder, *op.cit.*: 129–139.

and are responsible for labor problems on bases and for administering compensation; in these ways, they promote the smooth operation of U.S. military bases.

4.3 Distribution

The distribution of local hosts involved in a transaction would also seem to be an effective method for a government to control its vulnerability.[20] This is because if a government fears that domestic base transactions will be suspended, it has the option of transferring the transaction to another local host (government), thus reducing the effects of hold-up by any single local political actor.[21] Furthermore, if the transaction is distributed, when cooperating with local political actors and when engaging in base politics, the government is not only able to select the candidate that it considers to be the most suitable, but it can also switch the hosts depending on objectives and conditions. Moreover, implying to the other party of the transaction that it may be replaced in an *ad hoc* manner may also serve to improve the government's leverage during negotiations.

However, the local political actors will also prepare preventative measures to avoid this situation. That is, local actors that wish to maintain their own influence may seek to monopolize all base transactions. For example, the strength of the hold-up by a local actor might be proportional to the actor's provision of Japanese bases to the U.S. as a whole. The cases of Denmark and Japan provide just one example. First, from the Cold War period until today, the bases provided by the Danish government to the United States have all been located in Greenland. In this sense, the local actor, Greenland, provides 100% of the bases. Moreover, the base agreement only permits the establishment of bases in Greenland, so in this sense, the possibility of the bases being transferred to other regions on the Danish mainland in the future is zero.

By contrast, the bases Japan provided to the United States are scattered widely across the country. The provision rate of Okinawa Prefecture, in which many are located, is 70% (based on the areas of the bases and facilities exclusively for the use of the U.S. forces stationed in Japan). In addition, in its SOFA (Article 2), there are no limitations on the regions of Japan in which the U.S. may locate its bases. In other words, at least based on the SOFA, the Japanese government can transact with any local host (government) in Japan. Therefore, from the viewpoint of distribution, the effects of hold-up by Okinawa are not as great as those of Greenland. However,

---

[20] Hirschman, *op.cit.*: 31.

[21] This is also the same when viewed at the international level. For example, the United States also increased the number of host states in the post-Cold War period. Specifically, it increased the number of host states and tried to avoid being strategically dependent on a specific host state by expanding its access rights, including the use of the pre-integrated bases, access to ports for refueling and repairs, rights of air passage, and the temporary opening of facilities for use by the military of the host state. Harkavy, R. E. 2005. "Thinking About Basing," *Naval War College Review*, Vol.58, No. 3: 12–42.

if speaking only of Japan domestically, we understand that the effects for Okinawa Prefecture are larger than for any other local government (such as in Aomori Prefecture or Yamaguchi Prefecture).

5 Conclusion

Thus far, we have considered the conditions under which local political actors hold up governments to influence policy in base politics during peacetime in democratic states. From this, it would seem the following three hypotheses can be derived.

1. The influence of local political actors is determined by the vulnerability of the government during base transactions.
2. The government's vulnerability is expressed as a function of the (1) substitutability, (2) urgency, and (3) specificity of the base transacted.
3. The government can partially control the extent of its vulnerability by (1) integrating, (2) institutionalizing, and (3) distributing the transactions with local political actors.

These hypotheses were not clarified in previous studies focused on the political system of the host state and on the awareness of the policy decision-making elite. In other words, they bring to light the causal factors behind variations in democratic host states, specifically, of differences between stably operated bases and unstable bases as well as of differences between local governments that have influence over the central government and those that do not in the domestic political process of a specific host state. In other words, they can be said to suggest the possibility that these differences are created from the government's vulnerability, which is derived from the individual characteristics of the base (i.e., substitutability, urgency, and specificity), as well as from differences in the form of base transactions (i.e., integration, institutionalization, and distribution).

References

Calder, K. E. 2007. *Embattled garrisons: Comparative base politics and American globalism*, 116–119. Princeton: Princeton University Press.

Carroll, G. R. and D. J. Teece. 1999. *Firms, markets, and hierarchies: The transaction cost economics perspectives*. Oxford/New York: Oxford University Press.

Cooley, A. 2005. *Logics of hierarchy: The organization of empires, states, and military occupations*, 20–42. Ithaca: Cornell University Press.

Cooley, A. A. 2008. *Base politics: Democratic change and the U.S. military overseas*. Ithaca: Cornell University Press.

Friman, H. R. 1993. Side payments versus security cards: Domestic bargaining tactics in international economic negotiations. *International Organization* 47 (3): 387–410.

Goldmann, K. and G. Sjostedt. eds. 1979. *Power, capabilities, interdependence: Problems in the study of international influence*. London/Beverly Hills: Sage.

Johnson, C. 2004. *The sorrows of empire: Militarism, secrecy, and the end of the Republic*. New York: Metropolitan Books, Henry Holt.

Kawase, M. 2013. *Base maintenance policy and finances*, 2–3. Tokyo: Nihon Keizai Hyoronsha (In Japanese).

Keohane, R. O. and J. S. Nye. 2001. *Power and interdependence*. 3rd ed, 12. New York: Longman.

Knudsen, O. 1979. Multinational corporations and the small industrialized state: Some aspects of dependency and potential power. In *Power, capabilities, interdependence: Problems in the study of international influence*, ed. K. Goldmann and G. Sjostedt, 221–222. London/Beverly Hills: Sage.

Milgrom, P. and J. Roberts. 2003 *Organizational economics*, 44. Trans. Okuno, M. et al. Tokyo: NTT Publishing (In Japanese).

Sandars, C. T. 2000. *America's overseas garrisons: The leasehold empire*. New York: Oxford University Press.

Simmel, G. 2016. *Sociology: Inquiries into the construction of social forms* (Part 1), 198–200. Trans. Inui, M. Tokyo: Hakusuisya (In Japanese).

Stein, H. ed. 1963. *American civil-military decisions: A book of case studies*. Alabama: University of Alabama Press.

Teece, D. J. 1980. Economics of scope and the scope of the enterprise. *Journal of Economic Behavior Organization* 1 (3): 223–245.

Vine, D. 2009. *Island of shame: The secret history of the U.S. military base on Diego Garcia*. Princeton: Princeton University Press.

Williamson, O. E. 1975. *Markets and hierarchies: Analysis and antitrust implications*, 16–19. New York: Free Press.

Yamamoto, Y. 1989. *International Interdependence*, 111. Tokyo: University of Tokyo Press (In Japanese).

Yeo, A. 2011. *Activists, alliances, and anti-U.S. base protests*. New York: Cambridge University Press.

Greenland's Quest for Autonomy and the Political Dynamics Surrounding the Thule Air Base

Minori Takahashi

Abstract The principal goal of this chapter is to empirically and inductively answer the main topic of the volume, the influence of sub-state actors on national security, by focusing on the correlation between Greenland's quest for autonomy and the political dynamism regarding Thule Air Base. The important points the author wishes to make in this chapter are that during the negotiations on the radar upgrade and the transformation of Thule Air Base into an outpost in the U.S. missile defense shield Greenland on the surface maintained a negative stance towards the continuation of Thule Air Base while, in fact, arguing on the premise that the base would stay, and that, through the continued existence of the base and its transformation into a missile defense outpost, Greenland was attempting to increase its say in diplomatic and military matters. In contrast with that, to Denmark the Thule problem posed the question of neglect and responsibility - Denmark found itself in a situation where it had to take a visible, concrete approach toward Greenland and conduct negotiations while constantly trying to keep the political cost down. The attitude of Denmark to deal with the wishes of its sub-state actor Greenland with flexibility and compromise was a pervasive leitmotif in the discussions regarding the missile defense shield.

Keywords Greenland · Arctic · Autonomy · Thule Air Base · Missile defense · Indigenous people · Self-perception/self-image

---

M. Takahashi (✉)
Slavic-Eurasian Research Center and Arctic Research Center, Hokkaido University, Hokkaido, Japan
e-mail: minoritakahashi@slav.hokudai.ac.jp

M. Takahashi (ed.), *The Influence of Sub-state Actors on National Security*, Springer Polar Sciences, https://doi.org/10.1007/978-3-030-01677-7_3

1 Introduction

The purpose of this chapter is to empirically and inductively respond to the main theme of the present volume, the influence of sub-state actors on national security, by focusing on the correlation between Greenland's quest for autonomy and the political dynamics surrounding Thule Air Base.

As already mentioned in the introductory chapter, the discussions regarding the strengthening of the U.S. missile shield in 2000s provided the sub-state actor of Greenland with an opportunity to bolster its *de jure* involvement in the field of national security. If so, what was the logic by which Greenland was guided in using the base as its bargaining asset to fulfill its demands (or longing) for autonomy? And, in relation to that, with what intentions did Denmark swallow Greenland's demands and devolve its powers, while keeping the base in Thule as a venue for its own deal-making with the United States? Answering questions such as these is the purpose of this chapter.

I wish to point out the need to clearly distinguish between the aspirations for autonomy (the intention) and the acquisition of actual rights (the outcome). Here, rather than assessing the effectiveness or the degree of the influence Greenland managed to acquire, the author will seek to comprehend in an orderly manner, while utilizing parliamentary minutes and other material, what and under which circumstances Greenlanders were thinking and saying by focusing on their intention to demand further autonomy. Consequently, the present chapter will, based on the case of Greenland, show that aspirations for autonomy can exceed the realm of internal politics and grow to encompass international politics too.

2 The Construction and Operation of Thule Air Base

The U.S. strategic interest in Greenland has been embodied in the construction of military bases and meteorological stations throughout the island (DUPI 1997). On 9 April 1941 the United States concluded an agreement on the protection of Greenland with the then suzerain of Greenland, Denmark (or, to be exact, with the plenipotentiary Danish ambassador to Washington Henrik Kauffmann). In accordance with this "Agreement on the Protection of Greenland by the United States of America between the Plenipotentiary Ambassador to Washington Kauffmann and the United States Government",[1] the U.S. started building military bases on the

[1] Immediately after German forces invaded Denmark, Henrik Kauffmann, believing that "now that Denmark cannot function, it is my duty as a plenipotentiary ambassador to protect it," at his own discretion concluded the above agreement, permitting the stationing of U.S. troops in Greenland. Brun, Eske. "Grønlands Administration i Adskillelsens Aar". *Det Grønlandske Selskabs Aarsskrift 1946*. København: Ejnar Munksgaards Forlag: 36. Since concluding the agreement was Kauffmann's private decision, the document lost effect on the 12th of the same month, but was later ratified on 16 May 1945, after Denmark was liberated from Germany. Other than Kaufmann,

entire territory of Greenland on the premise of purely defending Greenland, without offensive capabilities.[2] Based on the Article 10 of the agreement, which clearly stipulated that "withdrawal is possible only when the existing threats to the peace and prosperity of the North American continent cease" and thus offered grounds for retaining the U.S. military presence (DUPI 1997: 52), the stationing of American troops was continued after the Second World War, too. In the 1950s, a plan to build the largest air force base in the Arctic in Thule, where until then only meteorological stations existed, was put forth (Berlingske Tidende 1949). However, in the area of Uummannaq-Dundas settlement at Thule, which was scheduled to be used for the base, hamlets had been formed, so the issue of what to do with the indigenous inhabitants became a major problem.

Denmark, which chose to continue its security relations with the U.S., on 27 April 1951 concluded with Washington the Agreement Relating to the Defense of Greenland (which went into effect on 8 June of the same year) for the purpose of building and maintaining NATO military bases, thereby allowing full-scale construction of military installations in Greenland. The construction work on Thule Air Base had started in March 1951, just before the signing of the agreement (Yoshitake 1985; Kawana 2012) . From that point onwards, the residents of Uummannaq were forced to live just next to the base, but, when the physical perimeter of the base was expanded in 1953 (Gad 2017a, b), they (26 households with 116 inhabitants)[3] were ordered by the Danish authorities to relocate (The order was transmitted to the residents on 25 May 1953) (Environmental Law Alliance Worldwide 2010).

It was in such context that the Thule base was constructed and started to operate. Since the second half of the 1950s the Distant Early Warning Line (DEW Line) was established as a string of forward outposts toward the Soviet Union (the Eastern block) (DUPI 1997: 339), strengthened by radar stations belonging to the Ballistic Missile Early Warning System (BMEWS) (Buchwald et al. n.d.), which included Thule, and was regarded as having strategic value. B-52 bombers had the mission of keeping an eye on the Soviet Union and protecting the radar sites. During one of such patrols, an accident occurred. It was 4.39 pm on January 21, 1968. A U.S. B-52 bomber crashed into the sea ice, 11 kilometers to the west of the Thule base. A fire broke inside the plane after the crew had turned on the heating, and six out of the seven members managed to eject to safety. However, it is thought that the four

---

the 1941 agreement was signed by the US Secretary of State Corden Hull. Forslag til Rigsdagsbeslutning i Anledning af den i Washington den 9. April 1941 / Forslag til Rigsdagsbeslutning i Anledning af den i Washington den 9. April 1941 undertegnede Overenskomst mellem Danmark og Amerikas Forenede Stater om Grønlands Forsvar. *Grønland under den kolde krig –Bilag*: 13–23.

[2] However, the functioning of Narsarsuaq, which served as a base for refueling on the air route back to Britain during counterattacks on Germany is often seen as an exception to this principle.

[3] According to a report by the international law expert Ole Spiermann, there were 27 households. Spiermann, Ole O. 2004, Hingitaq 53, Qajutaq Petersen, and Others v. Prime Minister's Office (Qaanaaq Municipality and Greenland Home Rule Government Intervening in Support of the Appellants)." The American Journal of International Law. Vol.98, No.3: 572–578.

nuclear bombs the plane was carrying broke, spreading radioactive substances over a radius of more than 10 kilometers.[4]

Let us now turn our attention to the political background of the accident. The coalition consisting of three parties: the Social-democratic Party (Socialdemokratiet), the Radical Left Party (Det Radikale Venstre) and the Justice Party of Denmark (Danmarks Retsforbund), which took power in May 1957, formulated a "nuclear policy" which strongly rejected the deployment of nuclear weapons on the Danish territory. What is more, the prime minister and foreign minister in the new government H. C. Hansen formally expressed opposition to the introduction of nuclear weapons to his country by rejecting the deployment of such weapons in his speech introducing government policies (DUPI 1997: 401, 544). And yet, despite that, it was later revealed that nuclear weapons had been brought to Greenland and, from 1958 to 1965, stationed in the Thule Air Base (United States Strategic Air Command 1979: 88–90; DUPI 1997: 331). That is, starting from February 1958, over the course of 8 months, 15 bombs not containing nuclear material and four surface-to-air nuclear-armed missiles "Nike" were brought to the base. Furthermore, from December 1959, as many as 48 nuclear warheads were transported to Thule. This practice continued until the summer of 1965, for, it was said, the purpose of ensuring the functioning of the air defense system of the base. This was made possible by the fact that when calls to conclude an agreement regarding the deployment of nuclear weapons to the Thule base were made, Hansen, who had supposedly put forward a clearly-defined "nuclear policy", avoided giving a clear-cut statement on whether his government would permit the stationing of such weapons.[5] The B-52 accident was, it may be said, an extension of that.

These events and the accident were later subsumed under the term "The Thule affair" and became one of the hottest issues in Danish post-Cold War politics. Thanks to several research projects (all of which were commissioned by the government under parliamentary pressure) by the Danish Institute of Foreign Policy (DUPI), which was founded as an independent institution in 1995, as well as the Danish Institute for International Studies (DIIS), which was established in 2003 and absorbed the DUPI, it is now clear that the structure of the Thule problem was shaped by the asymmetrical relationship between the U.S. and Denmark, where Greenland, which was left out of these considerations regarding national security, served only as the stage. However, as Greenland, from the status of a colony, got

[4]About the exposure of the clean-up workers to radiation, the way the accident was handled, the BBC report which claimed that one of the four nuclear warheads carried by the B-52 bomber that had been reported as recovered, was actually still missing, as well as later empirical research results and the lawsuit before the European Court of Human Rights.

[5]H.C. Hansen-papiret. Meddelelse (uformel) fra H.C. Hansen til den amerikanske ambassadør Val Peterson. 16. november 1957. UM 105.D.1.a. Hansen, in a private, unofficial letter dated 16 November 1957, regarded as top secret and later termed the "H.C. Hansen document" (H. C. Hansen-papiret), starts by emphasising that he is writing "on a strictly personal basis" and goes on to say: "I do not think that your remarks require any comments from my side (Jeg tror ikke, at Deres bemærkninger giver anledning til nogen kommentarer fra min side)" – an unclear position that can be interpreted both ways.

incorporated into the Danish state in 1953, as it further experienced modernization under the guidance of Denmark (in the 1950s and 60s) and the accession to the European Community (1973), as it obtained autonomy rights (1979) and then exercised them to exit the EC (1985), it, both in form and substance, progressively increased its self-reliance, so the gap in relation to the defense agreement of 1951, which was signed at the time Greenland was still a colony, became pronounced. Especially after the acquisition of autonomy, Greenland started making appeals to Copenhagen to be thorough in releasing relevant information, based on the Article XIII of the Home Rule Act of 1978, which stipulates that "treaties which require the assent of the Folketing and which particularly affect Greenland interests shall be referred to the home rule authorities for their comments before they are concluded". For example, when the upgrading of the Thule BMEWS became an issue in the 1980s, the then Greenland's Prime Minister Jonathan Motzfeldt and others voiced their displeasure with the amount of information on defense and security matters given to Greenland (Archer 2003). The Thule affair was the occasion which not only instilled Greenland with frustration towards the defense agreement, but also made it keenly aware of the necessity of its own engagement in the field of security (Or, if we borrow the wording of Chapter "Base Politics and the Hold-up Problem", we can say that the urgency of the base itself thereby increased).[6] As such trends were deepening, in December 2002 the Danish state received an official request from the U.S. for turning Thule Air Base into a missile defense outpost (that is, for upgrading the radar sites in the base).

3 Thule Air Base in the 2000s: Greenland's Home–Rule Parliament

The idea to turn Thule Air Base into a part of the U.S. missile defense shield became one of the items on the parliament agenda when, during the Clinton administration, the U.S. National Missile Defense plan was formulated and its application to the Thule base began to be explored in 1998 and 1999. In light of the Thule affair, Greenland watched the situation unfold with unease. However, at the same time a question emerged of how Greenlanders could exercise influence on the concerned part of their territory, so, on 18 November 1999, for example, a proposal was made in the parliament that, in case a request regarding missile defense is received, Greenland should make an appeal to Copenhagen and get directly involved in the negotiations.[7]

[6] Ulrik Pram Gad, personal message, 18 May 2017.

[7] Inatsisartuts samlinger Forårssamling 2003. 22. mødedag, onsdag den 9. april 2003, Dagsordenspunkt 13–1 (kl.13:59-). Hans Enoksen, Landsstyreformanden, Siumut. Inatsisartuts samlinger Efterårssamling 2002. 1. mødedag, fredag den 20. september 2002, Dagsordenspunkt 1 Åbningstale. Jonathan Motzfeldt, Landsstyreformanden, Siumut.

As the necessary footing for such participation, the revision of the 1951 defense agreement became the greatest subject of interest in the parliament.[8] Especially since the 50th anniversary of the signing of the agreement, which was concluded on 27 April 1951 and deemed "the root of all evil" (Information 2003) the calls for its renegotiation and revision significantly intensified.[9] MP Anthon Frederiksen pointed out that Greenlanders must not return to the past symbolized by the Thule affair and should face Copenhagen with a strong stance, while at the same time maintaining good cooperation with it.[10] Kuupik Kleist appealed for the revision of the defense agreement by saying that Greenland had a duty as a constituent of the Danish state, and should, within that framework, create a new mechanism so that the Cold War incident would not be repeated.[11] Concrete expressions of these voices were conveyed on 8 November 2001 to the then Danish Minister of Foreign Affairs Mogens Lykketoft, and on 11 January 2002 to his successor Per Stig Møller. Nonetheless, the Danish government kept its stance that the revision of the defense agreement was not necessary despite such requests from Greenland, and no breakthrough in the talks was achieved. However, progress was made on 30–31 January 2002 during the summit between the then Danish Prime Minister Anders Fogh Rasmussen and the then Greenland's Premier Jonathan Motzfeldt. At the meeting, where the impact of the defense agreement on Greenland was discussed, the Danish government did not accept the revision of the agreement, viewing it as the optimal and sustainable framework for promoting cooperation between the U.S., Denmark and Greenland. Instead, it was agreed that the contribution to the solution of problems burdening Greenland should be sought through a supplemental agreement.[12] Having this background in mind, MP Jensine Berthelsen, for example, called for caution in preparations for the answer about whether to link the revision of the defense agreement to the missile defense shield or not, in view of the fact that the two were of different origin.[13] After the U.S. official request came in December 2002, the issues of how relating the revision of the agreement to the missile defense should be understood and what should be done were laid on the table for discussion in more concrete terms.

---

[8] Inatsisartuts samlinger Forårssamling 2003. 6. mødedag, tirsdag den 11. marts 2003, Punkt12, Udenrigspolitisk Redegørelse. Hans Enoksen, Landsstyreformand, Siumut.

[9] Inatsisartuts samlinger Efterårssamling 2001. 02FM/01.25.01–35 Beslutningsforslag om ændring af den dansk-amerikanske forsvarsaftale af 1951. Fra Maliinannguaq Markussen Mølgaard, Inuit Ataqatigiit. d. 20. august 2001.

[10] Inatsisartuts samlinger Forårssamling 2003. 7. mødedag, onsdag den 12. marts 2003, kl.13:00–18:00. Anthon Frederiksen, ordfører, Kandidatforbundet.

[11] Inatsisartuts samlinger Forårssamling 2003. 6. mødedag, tirsdag den 11. marts 2003, Punkt12, Udenrigspolitisk Redegørelse. Kuupik Kleist, ordfører, Inuit Ataqatigiit.

[12] Inatsisartuts samlinger Forårssamling 2002 Svarnotat 2. beh. (Efterårssamling 2001. 02FM/01.25.01–35 Beslutningsforslag om ændring af den dansk-amerikanske forsvarsaftale af 1951. Fra Maliinannguaq Markussen Mølgaard, Inuit Ataqatigiit. d. 20. august 2001).

[13] Inatsisartuts samlinger Forårssamling 2003. 7. mødedag, onsdag den 12. marts 2003, kl.13:00–18:00. Jensigne Berthelsen, ordfører, Atassut.

4 Greenland's "Refusal"

The then Prime Minister Hans Enoksen stressed that Greenland's commitment to the polar, Nordic and European, as well as global cooperation was conducive to achieving Greenland's vital interests[14] and that intense involvement of Greenland in the discussions regarding the missile defense and the negotiations on the defense agreement was necessary, as debates on such matters should not be exhausted only in the talks between Copenhagen and Washington.[15] Also, in light of the Thule affair, there was a shared conviction in the parliament, that if Greenland could communicate directly with Washington, the U.S. would show more respect for it. Behind this was the belief that Copenhagen was not particularly keen on sharing information and empowering Nuuk to get involved as a negotiator.[16] Furthermore, based on the understanding that world-wide trends regarding security after 9.11 would, directly or indirectly, inevitably affect Greenland, there was a shared opinion in the parliament that Greenland needed to consider what stance it should take towards the war on terror and make its thinking in that respect known.[17] Such view, for example, can be seen in Aqqaluk Lynge, Jørgen Wæver-Johansen and others.[18] Thus, from the beginning an understanding existed that the most important thing was that Greenland itself committed to the negotiations and influenced some of the decisions.

Greenland's next move was a refusal.[19] This was because in Greenland's parliament an understanding prevailed that consent could not be given since it was not clear what the incorporation of Thule Air Base in the missile defense shield (the upgrading of radar installations) actually meant and what risks might be involved. This, of course, can be taken literally as Greenland's politically principled rejection of the missile defense function and its implementation in Thule. Corroborating such interpretation is the fact that the functioning of the missile shield itself and the negative effects it might bring were often taken up for discussion in Greenland's parliament. For example, Jensine Berthelsen argued that the rejection of the strengthening of the U.S. missile defense in Greenland was beyond objection since improving and extending the radar network would badly affect the world peace by contributing to a new arms race (Dragsdahl 2003).

[14] Inatsisartuts samlinger Forårssamling 2003. 6. mødedag, tirsdag den 11. marts 2003, Punkt12, Udenrigspolitisk Redegørelse. Hans Enoksen, Landsstyreformand, Siumut.

[15] Inatsisartuts samlinger Forårssamling 2003. 7. mødedag, onsdag den 12. marts 2003, kl.13:00–18:00. Hans Enoksen, Landsstyreformand, Siumut.

[16] Ulrik Pram Gad, personal message, 18 May 2017.

[17] Inatsisartuts samlinger Forårssamling 2003. 6. mødedag, tirsdag den 11. marts 2003, Punkt12, Udenrigspolitisk Redegørelse. Hans Enoksen, Landsstyreformand, Siumut.

[18] Inatsisartuts samlinger Forårssamling 2003. 1. mødedag, tirsdag den 4. marts 2003, kl.13:00–18:00. Aqqaluk Lynge, ordfører, Inuit Ataqatigiit. Inatsisartuts samlinger Forårssamling 2003. 6. mødedag, tirsdag den 11. marts 2003, Punkt12, Udenrigspolitisk Redegørelse. Jørgen Wæver-Johansen, ordfører, Siumut.

[19] Inatsisartuts samlinger Forårssamling 2003. 7. mødedag, onsdag den 12. marts 2003, kl.13:00–18:00. Hans Enoksen, Landsstyreformanden, Siumut.

However, as Jørgen Dragsdahl points out from the viewpoint of peace studies, there is room for the interpretation that Greenland's refusal was not an expression of a political principle, but rather, a negotiation card (Dragsdahl 2005a). For example, Per Berthelsen admits that the act of rejection was not meant to close the door to dialogue, but was aimed to challenge the assumptions made by Denmark and the U.S., as the other side in the negotiations.[20] If we accept that in Greenland the rejection was commonly perceived and shaped as a card for negotiation, what was the plan behind it? Judging by the parliament minutes, we may say that there were at least two objectives.[21] One was to effect a change in the still valid 1951 Agreement Relating to the Defense of Greenland,[22] and the other was to give a concrete shape to the environment clean up, in light of the facts that plutonium contamination[23]due to the B-52 crash was affecting a portion of Thule, and that internal contamination by radioactive cesium, caused by the 1986 nuclear accident in Chernobyl, was detected among the native Sami population (Ackrén and Jakobsen 2015).

Greenlandic MP Kuupik Kleist, based on these two points, proposed a process consisting of the following steps: 1. the renegotiation of the 1951 defense agreement with Denmark, which serves as the departure point for everything else, 2. presenting the U.S. with the results of those negotiations regarding *not only military matters but also environmental cooperation*, and 3. ultimately linking all that with the amendment of the entire defense agreement, and the application of the newly agreed terms to the case of the incorporation of the Thule base into the missile defense shield.[24] In other words, as can be seen from, Per Berthelsen's words, Greenland, clearly seeing the U.S. request regarding the missile defense as an opportunity to renegotiate the 1951 defense agreement,[25] was planning to not let the talks with the U.S. be confined only to military matters, and to have its further demands, such as the solution of the environmental problems, incorporated.

Greenland's intent to develop diverse cooperation via the base was resting on maintaining the high specificity that military bases in Polar Regions possess, which

---

[20] Inatsisartuts samlinger Forårssamling 2003. 22. mødedag, onsdag den 9. april 2003, Dagsordenspunkt 13–1 (kl.13:59-). Per Berthelsen, ordfører, Demokraterne.

[21] Inatsisartuts samlinger Forårssamling 2003. 7. mødedag, onsdag den 12. marts 2003, kl.13:00–18:00. Kuupik Kleist, ordfører, Inuit Ataqatigiit.

[22] Enoksen was clearly talking not about a partial modification but "a new agreement", which can be taken to mean that he was aspiring for the formulation of an entirely different agreement or for a large revision. Inatsisartuts samlinger Forårssamling 2003. 6. mødedag, tirsdag den 11. marts 2003, Punkt12, Udenrigspolitisk Redegørelse. Hans Enoksen, Landsstyreformand, Siumut.

[23] Inatsisartuts samlinger Efterårssamling 2002. 02EM/01.25.01–96 Spørgsmål til Landsstyret om miljøundersøgelser af det forurenede område omkring Thule Air Base. Fra Mads Peter Grønvold. den 30. juli 2002. Inatsisartuts samlinger Efterårssamling 2002. Svarnotat, den 30. september 2002 (02EM/01.25.01–96 Spørgsmål til Landsstyret om miljøundersøgelser af det forurenede område omkring Thule Air Base. Fra Mads Peter Grønvold. den 30. juli 2002).

[24] Inatsisartuts samlinger Forårssamling 2003. 6. mødedag, tirsdag den 11. marts 2003, Punkt12, Udenrigspolitisk Redegørelse. Kuupik Kleist, ordfører, Inuit Ataqatigiit.

[25] Inatsisartuts samlinger Forårssamling 2003. 6. mødedag, tirsdag den 11. marts 2003, Punkt12, Udenrigspolitisk Redegørelse. Per Berthelsen, ordfdører, Demokraterne.

lies in the fact that they are physically difficult to move, while at the same time seeking an extra value for the base in fields different from its distinctive purpose (i.e., military, security use). Indeed, much interest was shown for the operation of the Thule base in the context of the changes in the specificity of military bases in the post-Cold War period, such the fall in the incentive to restrict their role purely to military purposes (Berlingske Tidende 2001). In that period Greenland's actions aimed at diversifying the role of the base in Thule were meaningful and were received as such.

5 The Inclination Towards the Continuation of the Base and the Expansion of Greenland's Powers

Upon the reception of the official U.S. request, Prime Minister Hans Enoksen, although speaking of refusal, was actually trying to discern the extent of Greenland's possible involvement in the negotiations regarding the updating of the defense agreement. At the same time, he was openly advocating the need to discuss the content of a "compensation package". That is, his thinking was that the negotiations about the missile defense shield were a hard sell if they were to remain limited *only* to the renegotiation of the defense agreement and its updating.[26] The important points to note here are that, consciously or not, on the surface he maintained a negative stance towards the continuation of the Thule base while in fact arguing on the premise that the base will stay, and that, through the continued existence of the base and its transformation into a missile defense outpost, he was attempting to increase Greenland's say in diplomatic and military matters.[27] This assumption that the base will remain can also be seen in the stance of the home-rule parliament emerging as a conclusion in the last of the following four points debated in that legislative body[28]:

- Impact on the environment

What would be the impact of the strengthening of the base on the natural environment (considering that in the northwest of the island, where Thule Air Base is located, humans are particularly closely tied to the environment)? Regarding this, it was pointed out that the American request referred to *the upgrading of the already existing radars*, and did not imply the installation of new facilities or functions. Nor did it necessarily mean that missiles would be brought to Greenland and stationed there. For that reason, the delegates were of the opinion that the question of the

[26] Inatsisartuts samlinger Forårssamling 2003. 22. mødedag, onsdag den 9. april 2003, Dagsordenspunkt 13–1 (kl.13:59-). Hans Enoksen, Landsstyreformanden, Siumut.

[27] Inatsisartuts samlinger Forårssamling 2003. 6. mødedag, tirsdag den 11. marts 2003, Punkt12, Udenrigspolitisk Redegørelse. Hans Enoksen, Landsstyreformand, Siumut.

[28] Inatsisartuts samlinger Forårssamling 2003. 22. mødedag, onsdag den 9. april 2003, Dagsordenspunkt 13–1 (kl.13:59-). Hans Enoksen, Landsstyreformanden, Siumut.

burden of the radar upgrade on the environment had to be low on the list of priorities to be considered.

- Impact on the economy

Regardless of whether Greenland accepts the U.S. request or rejects it, for the U.S. the continuation of the base is a high priority. If so, then the choice of accepting the upgrade and considering what benefits can be thereby gained, especially what economic effect of having the base continue to operate could be achieved, also exits. Regarding this, an estimate was mentioned that, at that point, the potential economic effect, such as tax revenue, etc., could amount to approximately ten million Kroner per year.

- Contribution to collective security

Through Thule a contribution to the security of Greenland, Denmark and NATO is made and if the upgrade of the radar installations is realized, that contribution can be further enhanced. When it comes to long-term guarantees for security, cooperation with NATO will help Greenland carry out its duties as a member of the international community and will in the long run be beneficial for the enhancement of its autonomy and for the entire society.[29]

- Refusal as a choice

In the light of its recent commitment to the war in Iraq it is conceivable that Denmark will, as an expression of good will, support the request by the United States even if Greenland announces its refusal. That is what Denmark has done in the past, and that is what has resulted in Greenland's experiencing "dark historic times". Based on the premise that such events should not be repeated in the future, Greenland must make its influence felt through the medium of joint responsibility.

6 The Response of Denmark

Meanwhile, in Copenhagen, all political parties, except the Danish People's Party and the United List, were in principle in favor of furthering cooperation with Greenland within the framework of the Danish constitution (i.e., within the

[29] To borrow Enoksen's words, what was important for the realization of Greenland's aspiration to take a chair at the table negotiating the 1951 defense agreement was the change in the U.S. understanding of that agreement. The U.S., although stressing that the agreement was the optimal framework for the operation of Thule Air Base, recognized the need for its amendment, i.e., for the amendment of the way Greenland had been treated since 1951. Inatsisartuts samlinger Forårssamling 2002 Svarnotat 2. beh. (Efterårssamling 2001. 02FM/01.25.01–35 Beslutningsforslag om ændring af den dansk-amerikanske forsvarsaftale af 1951. Fra Maliinannguaq Markussen Mølgaard, Inuit Ataqatigiit. d. 20. august 2001). Inatsisartuts samlinger Forårssamling 2003. 6. mødedag, tirsdag den 11. marts 2003, Punkt 12, Udenrigspolitisk Redegørelse. Hans Enoksen, Landsstyreformand, Siumut.

framework that enables the containment of the risk that Thule Air Base could be held-up by Greenland and certain control of Denmark's vulnerability regarding the U.S. base in Greenland). Debates were conducted on the premise that Greenland's viewpoint, that is, its demand for the revision of the 1951 defense agreement and strong wish to *enhance its cooperation with the U.S.*, would be reflected in Denmark's decision regarding the reply to the U.S..[30] For example, Prime Minister Anders Fogh Rasmussen made a remark that the basic stance of Denmark was that the reply to the U.S. should be given based on a mutual understanding with Greenland.[31] MP Jeppe Kofod stated that the 1951 agreement, which was concluded during the colonial period, made no sense in the present day, and that the two sides would soon reach a positive conclusion.[32] Furthermore, Gitte Seeberg expressed the opinion that the decision to be made should reflect the new times and the new relationship, and that, in order to confer benefit on the people of Greenland, a multifaceted agreement with a content broader than military matters should be concluded.[33] These statements were based on the understanding that the improvement of the base should not just contribute to the security of the United States, but also to the security of the entire *Danish state*, that is, on the idea that that the arrangements should be beneficial for all three parties.[34] For that reason, it was thought that the final decision needed to be an agreement between the U.S., Denmark and Greenland, that its content should not be restricted to military matters but be broader in scope, and that Greenland should be actively involved as one of its signatories.[35] The important point was how to concretely incorporate Greenland's voice in the policy making decisions, and not as just as a programmatic idea, and how to make Greenlanders feel the difference.

That being said, as can be seen in Jens Hald Madsen's debate, here and there doubt was also expressed in the sense that, within the framework of the Danish state (the Danish constitution), Denmark was already making substantial efforts to engage in close dialogue with Greenland (Broholt n.d.), and that it was not clear what Greenlanders meant by "exercising influence". Also, Peter Skaarup and other MPs from the Danish People's Party expressed the view that, although they desired joint participation of Greenland and Denmark in the decision making, it should be emphasized that, in accordance with the Article 19 of the constitution, the right to make the final decision rested solely with Denmark, and that because of that it was clear that Denmark could conduct negotiations with the U.S. and make the decision

[30] Inatsisartuts samlinger Forårssamling 2003. 22. mødedag, onsdag den 9. april 2003, Dagsordenspunkt 13–1 (kl.13:59-). Hans Enoksen, Landsstyreformanden, Siumut.

[31] Forespørgsel nr.F29: Forespørgel til statsministeren og udenrigsministeren: Hvad kan statsministeren og udenrigsministeren oplyse om den danske regerings holdning til missilskjoldprojektet, herunder til den amerikanske regerings anmodning om at inddrage Thuleradaren i forbindelse med missilskjoldprojektet? Den 29. april 2003. Kl.15:40.

[32] Ibid. Kl.16:45.

[33] Ibid. Kl.17:45.

[34] Ibid. Kl.15:45.

[35] Ibid. Kl.15:50.

on its own if it wanted to do so, but that, nonetheless, it wished to act in unison with Greenland.[36]

7 Concern for the Residents of Qaanaaq

As, in such circumstances, the revision of the 1951 defense agreement was taking shape in coordination with Denmark, what constituted the grounds for Greenland's demands for more powers were the existence of the above-mentioned Thule problem and the rights of the inhabitants of Qaanaaq (the indigenous Inughuit people) as subjects under rule. Placing the history and disputes regarding the Thule problem, which fully came to light only after the end of the Cold War, on the agenda of the negotiations about the missile defense shield and making Denmark aware of them, was significant for Greenland in the sense that, in that way, legitimacy was given to its demands. Furthermore, the concern for the Qaanaaq residents, and references to the forced removal experienced by their earlier generations and the history of the usurpation of their hunting and fishing grounds, were some of the issues the Greenlandic side particularly had on its mind while pursuing negotiations with Denmark regarding the missile defense.[37] In other words, it may be said, perhaps, that, had Greenland's home-rule government pushed forward discussions regarding the missile defense shield without the residents of Qaanaaq or the collaboration with the Qaanaaq community, the reply given in such a way would have been inappropriate, both from the viewpoint of internal political dynamics in Greenland and in moral terms.[38] For example, Johan Lund Olsen, when talking about the Qaanaaq Inughuit residents who lost their living environment, their hunting and fishing grounds due to the forced removal (Lynge 2002), suggested that time for careful and sustained contemplation of the 1951 agreement and the missile shield, as well as time for coming to terms with all that was necessary, considering the residents' growing concern that Thule might become a target of a nuclear attack after the transformation of the base into a missile defense post.[39] Since, in the worldview of the Inughuit, the coexistence of humans and animals is seen as an inseparable "social relationship", being deprived of land for them meant not just the loss of sustenance but of the worldview, too.[40] Thus, it was necessary to include that world-

[36] Forespørgsel nr.F29. Op. cit. Kl.17:30.

[37] Inatsisartuts samlinger Forårssamling 2003. 6. mødedag, tirsdag den 11. marts 2003, Punkt12, Udenrigspolitisk Redegørelse. Inatsisartuts samlinger Forårssamling 2003. 7. mødedag, onsdag den 12. marts 2003, kl.13:00–18:00. Inatsisartuts samlinger Forårssamling 2003. 22. mødedag, onsdag den 9. april 2003, Dagsordenspunkt 13–1 (kl.13:59-).

[38] Inatsisartuts samlinger Forårssamling 2003. 22. mødedag, onsdag den 9. april 2003, Dagsordenspunkt 13–1 (kl.13:59-). Kuupik Kleist, Inuit Ataqatigiit.

[39] Inatsisartuts samlinger Forårssamling 2003. 7. mødedag, onsdag den 12. marts 2003, kl.13:00–18:00. Johan Lund Olsen, ordfører, Inuit Ataqatigiit.

[40] Here it should be noted that hunting and fishing can be viewed not only as concrete practices, but also as means of representing the identity of one's ethnic group.

view in thinking about the question of Greenland's self-determination in relation to the issue of U.S. missile defense.

Regarding that, there is a need here to take into consideration the multi-layered structure of the political dynamics that surrounds the U.S. base in Greenland that is not visible based solely on the trilateral relationship presented in Chapter "Base Politics and the Hold-up Problem". That is, if we assume that concern for Qaanaaq was underpinning Nuuk's assertions, then that suggests that another substratum, i.e., the dynamism of a local-local relationship (between Nuuk and Qaanaaq), may have functioned as an effective explanatory variable constituting the politics regarding the Thule base. We, thus, come to the understanding that Nuuk was trying to redefine the relationship between the actors (i.e., to bring about a revision of the defense argument) by skilfully articulating in the negotiation process the metamorphosis in the positionality of Greenland that arose from the incorporation of this local-local relationship (Gad 2017a).

In fact, it can be said that the rights of Greenlanders as "indigenous people" were the leverage in the negotiations. The reason I placed the term indigenous people under quote marks is that what it is to be an indigenous people involves the criterion of self-perception, that seems to contain defining elements such as "indigenousness",[41] "the status of the ruled",[42] "historical continuity"[43] and "self-identification[44]". In that sense, indigenous ethnic groups do not just "exist as such", but also "become such".[45] In the sense that, as Sachiko Kubota aptly points out, indigenous people "are able to claim distinctiveness in opposition to universality, with a high possibility that such distinctiveness will be recognized", (Kubota 2009) and that the recognition does not refer only to an official recognition (of their rights and status by a state) but also implies a revitalization (of the human-environment cycle) and reconciliation (though other forms in which indigenous people seek their rights, and different agreements such as those regarding land ownership and education), it has been pointed out that the concept of indigenous people, as a framework for self-assertion, is to a certain extent effective in terms of influencing outcomes when claiming rights.

We may say that behind Greenland's assertions regarding the missile defense there was an intention to demand further rights form the standpoint of an indigenous people. That was an extremely important constituent part of Greenland's moral

[41] They are groups of people or the descendants of groups of people who originally inhabited the areas occupied by Europeans at the time European powers were making incursions into "the New World", or who were forcibly removed from their original living environment due to such incursions.

[42] They are groups of people or the descendants of groups of people who are not able to enjoy their original life style due to a colonial status, or some other inferior social or legal status.

[43] A historical continuity exists in the sense that they are the descendants of the original inhabitants.

[44] Groups and individual members of groups who regard themselves as indigenous people.

[45] Behind this view is the notion that, aside from the "visible indigenous people" who have been recognized as such, there are also "latent indigenous people", who have not yet been recognized as indigenous or have the potential to become visible in the future.

authority, which Kristian Søby Kristensen regards as a source of power that helped raise the quality of Greenland's influence in its attempts to advance the negotiations in its favour (Kristensen 2004: 9). As Kuupik Kleist remarked, what was important in the negotiations regarding the missile defense shield was "how Denmark treated the indigenous people (i.e., Greenlanders) within its state".[46]

Indeed, to Denmark the Thule problem posed the question of neglect and responsibility (Tamnes and Holtsmark 2014). Furthermore, the end of the United Nations Third Decade to Combat Racism and Racial Discrimination, which had opened with the UN International Year of the World's Indigenous People in 1993, was fast approaching, so Denmark found itself in a situation where it had to take a visible and concrete approach toward Greenland and conduct negotiations while trying to keep the political cost down,[47] i.e., while maintaining 1. the territorial integrity of the Danish state and 2. the balance in securing political legitimacy throughout the territory by acquiring Greenlanders' support for its political authority.[48] According to scholars Svend Aage Christensen and Kristian Søby Kristensen, since Denmark sees itself as a democratic and moral country and Qaanaaq's past is an inconvenient truth (Christensen and Kristensen 2009) for it in terms of its self-perception, Denmark's political goal in facing that troublesome past was to try to repair its self-image and continue the cooperation with the U.S. in matters of national security without provoking Greenland's protest. That this self-image and the reputation that constitutes it have played an important role in the interaction between Denmark and Greenland is the argument of DIIS Senior Researcher Mikkel Runge Olesen, (Olsen 2017) and we may say that Denmark strongly felt the need to face its troublesome past in its attempts to win stable appraisal (i.e., trust) in the trilateral relations with the U.S. and Greenland.

8 The Itilleq Declaration

What Copenhagen, honoring Greenland's wishes, prepared as a concrete outcome, was a declaration (widely known as the Itilleq Declaration) reaffirming Greenland's *de jure* involvement in the sphere of national security within the framework of the Danish constitution.[49] I say reaffirming *de jure* involvement, because Greenland had

[46] Inatsisartuts samlinger Forårssamling 2003. 7. mødedag, onsdag den 12. marts 2003, kl.13:00–18:00. Kuupik Kleist, ordfører, Inuit Ataqatigiit.

[47] The coalition of the Liberal Party and the Conservative Party, formed in November 2001, was especially determined to promote the politics of consideration and respect for Greenland. See Archer, C.: 139.

[48] We take the view that when the gap between these two items is small, that is, when the central government can achieve both of the above two goals without any conditions attached, the political cost is low, while, on the other hand, the political cost is high when the realization of those goals is accompanied by the exertion of some sort of coercive force or influence.

[49] Joint statement of the Danish government and Greenland's self-rule government on the participation of Greenland in diplomatic and security policies. May 14, 2003.

already had the experience of *de facto* involvement in matters of national security. It is thought that at the beginning of the 1990s Greenland influenced Copenhagen's policy regarding the agreement signed on 13 March 1991[50] concerning the use of the Søndre Strømfjords military base (the present day Kangerlussuaq airport) and Kulusuk airport.[51] This was based on Article XIII of the Home Rule Act guaranteeing Greenland's position, which stipulates that "treaties which require the assent of the Folketing and which particularly affect Greenland interests shall be referred to the home rule authorities for their comments before they are concluded", and Clause 2 of the Article XVI which states that "the central authorities may after negotiations with the home rule authorities empower the home rule authorities to advance special Greenland interests by taking part in international negotiations of special importance for Greenland's commercial life". Adam Worm from the Representative Office of the Greenland's Self-Rule Government in Copenhagen has said that, considering the already existing achievements in exercising *de jure* influence, the 2003 declaration should be understood as reaffirming that influence.[52] In that sense, Denmark's response did not exceed the scope of a minor change. Nonetheless, it may be said that Denmark managed to answer the calls by Greenland, which aspired for greater autonomy and participation in the negotiations, in the sense that it made Greenland's involvement in the security sphere more visible. That is, we may interpret the Itilleq Declaration as Copenhagen giving Greenland the right to have a say in matters concerning the Thule base to the extent that Copenhagen is not held up.[53]

The declaration was interpreted as a necessary pre-condition for the negotiations about the missile defense shield and the conclusion of a future agreement that would contain a revision of the old agreement regulating defense issues.[54] In the declaration, Greenland is regarded as an actor that can "demand international negotiations" regarding problems important to it, as well as "participate" in them and "influence" them "on an equal footing".[55] This meant that Greenland was now able to have, to a certain extent, an influence on the international negotiations regarding the Thule Air

[50] Aftalememorandum mellem Kongeriget Danmarks Regering (Inklusive Grønlands Landsstyre) og Amerikas Forenede Staters Regering Vedrørende brug af Søndre Strømfjords Luftfartsanlæg, Kulusuk Flyveplads og Andre Forhold i forbindelse med de Forenede Staters Militære Aktiviteter i Grønland. Lovtidende 1991. København: udenrigsministeriet (den 19. september 1991).

[51] Interview with Adam Worm, 31 July 2017 at the Representative Office of the Greenland Home-Rule Government in Copenhagen.

[52] Ibid.

[53] The declaration was not signed in the Qaanaaq area where the Thule base is located, nor in the capital of Greenland Nuuk or the capital of Denmark Copenhagen, but in the village of Itilleq, which is where Greenland's Prime Minister Hans Enoksen was born. It is believed that this was a strategic choice in Denmark's attempts to give a concrete shape to its respect for Greenland's intentions. See Dragsdahl 2005b.

[54] Interview with Adam Worm on 31 July 2017 at the representative office of the Greenland home-rule government in Copenhagen.

[55] Forespørgsel nr.F60: Hvad kan regeringen oplyse om status i forhandlingerne mellem på den ene side Danmark og Grønland og på den anden side USA vedrørende en eventuel opgradering af Thule-radaren?. Den 26. maj 2004.

Base. And, as far as it can be seen from parliamentary minutes, that was the result of Greenland's efforts to gain more say in matters of security, while not attempting to shake up the framework of the Danish state itself.

Here I shall list the main points of the declaration:

1. Greenland can participate and influence diplomatic and security matters that are important to it,
2. Denmark and its internal autonomous territory of Greenland can participate and influence matters on equal terms in all issues affecting them,
3. Greenland's request for international negotiations regarding matters that are of special importance to it is a natural departure point. Greenland may participate in those negotiations together with Denmark and express its views.
4. Greenland can be, together with Denmark, one of the signatories of agreements based on the International Law.
5. The joint statement does not imply any pre-judgement or partiality in the process of Greenland's negotiations with external actors and will not lose effect until "multiple issues"[56] have been solved.

It is worth repeating here that Greenland is an autonomous territory of the Danish state (Danish Realm), which centers on mainland Denmark. According to the Article XIX of the Danish constitution, Greenland does not possess the right to make decisions regarding foreign affairs. The right to such decisions is given to Denmark (strictly speaking, to its government), and, thus, it should be borne in mind that whether Greenland can exercise the aforementioned rights depends on the political situation and Denmark's judgement. However, on the other hand, if we take into consideration that the declaration was formulated with a view to the negotiations on the missile defense shield and the renegotiation of the defense agreement that regulates it, we may say that the concrete influence Greenland acquired was accepted relatively positively.[57] If we assume that Greenland's demands for the renegotiation and the revision of the 1951 defense agreement have indeed resulted in the revision of that agreement, then we may say that, for Greenland, this minor change or supplementation of the agreement has a meaning that exceeds the amendment itself, even without resorting to quoting Mikiko Iwasaki's remark that "Even if the final legislative powers rest with the central government, if the voice of local governments can be reflected in the bills in the process leading up to the enactment of laws, than local governments can in effect become decision makers themselves" (Iwasaki 1998). Thus, we can now understand how Greenland, despite not being able to directly participate in the outcomes, but having the option of influencing the proceedings leading to those outcomes, has come to be viewed as an actor that both regulates its intentions (self-control) and concretely expresses those intentions

[56] The issues referred to here are: 1. the question of the missile defense, 2. the upgrade of the Thule base, 3. the updating of the 1951 defense agreement in relation to that and 4. technical and economic (financial) cooperation with the US and others relating to the fields of diplomacy and security.

[57] Forespørgsel nr.F29. Op. cit. Kl.18:20.

through actions (self-rule) by itself in a field that traditionally belongs to the jurisdiction of the state, and not as an internal constituent of the state that simply keeps agreeing or disagreeing with the center. As already pointed out in the 2008 report "Arctic in the Time of Change", Greenland's autonomy rights forged in the period between 2002 and 2004 and the process behind them are the most important elements in the progression to the conclusion of a complex, multilateral agreement (the Igaliku Agreement of 2004), and the negotiations with the U.S. regarding Thule Air Base are highly valuable in terms of understanding how Greenland's cooperation with Denmark developed thus far (Namminersornerullutik Oqartussat et al. 2008: 10).

On the other hand, for Denmark, the Itilleq Declaration was a result of a political process in which a response was given to the U.S. request based on a compromise and consensus with the sub-state entity of Greenland. Of course, at the root of this was the desire to peacefully obtain Greenland's consent and conclude a constructive agreement with the United States that would contribute to the interests of the Danish state (i.e.,Copenhagen's interests). In that sense Villy Søvndal's remark that things were moving according to a rigged scenario is spot on.[58] Also, if we take NATO to be an organization in which "countries with a high military contribution wield strong influence", (Ueda 2007) then using Greenland as the space for bargaining (Taagholt and Hansen 2001) with the U.S. and extending the validity of the Greenland diplomatic card was a choice of high priority for Denmark. The then Danish Prime Minister Anders Fogh Rasmussen clearly revealed his position when answering the question "Who more than anybody guarantees Denmark's security?": "My answer is as clear as the sky. The most stable maintenance of Denmark's security is achieved not within the vulnerable and unbalanced power relation between Germany, France and Great Britain, but by the super power, the USA (Berlingske Tidende 2004)."

Perhaps we need not go so far as to point out Stig Møller's remark that he realized the negotiations on the revision of the 1951 defense agreement in cooperation with Jonathan Motzfeldt, received the U.S. formal request in cooperation with Josef Motzfeldt and discussed how the two should be linked and dealt with in cooperation with Hans Enoksen[59] to bring home the fact that joint coordination and cooperation are a convincing explanatory variable for understanding Danish politics, where answers to problems are sought while maintaining political functioning and stability. Although we cannot literally apply the opinion of Kenneth E. Miller from the 1960s that if a political party (in Denmark) is branded as inflexible or uncooperative, it loses the next elections to the present-day politics (Miller 1968), we can say that the attitude of Copenhagen, as the only power center, to deal with the wishes of its sub-national entity Greenland with flexibility and compromise was a pervasive leitmotif in the discussions regarding the missile defense shield. That is a part of the political culture that has in the past often been explained with key terms such as "consultation" and "participation", and was, during the deliberations regarding the

[58] Forespørgsel nr.F29. Op. cit. Kl.16:15.

[59] Forespørgsel nr.F29. Op. cit. Kl.18:35.

missile defense shield, referred to as "the culture of consultation" and "the right to be heard" (Broholt n.d.). We may at least say that Denmark's political stance of aspiring for the formulation of an answer to the U.S. request while keeping down the political cost by creating an atmosphere of cooperation was visible.

In fact, this Danish political culture may have had a significant impact on the U.S. discourse and conduct. The details will be revealed in the next chapter - here I wish to point to the statement of Adam Worm from the Representative Office of Greenland in Copenhagen, according to whom, the U.S. showed respect for the Danish culture of consultation and thanks to that was able to gain trust from Denmark and Greenland.[60] Also, researcher Mikkel Broholt, quoting Nikolaj Petersen's comments, notes that the upgrade of the Thule base radars could have been realized even without Denmark's and Greenland's consent, and explains that the reason why the U.S. did not do so was that it chose not to ignore the Danish culture of consultation (Broholt n.d.: 4–5). Furthermore, at the time of the negotiations, Denmark's MP Jens Hald Madsen stated that the more U.S. learned about the history of the Qaanaaq region, the more it understood it could not avoid collaboration with Greenland. He also added that the U.S. was showing the will to engage in the work for finding solutions together with Denmark and Greenland, and that it was conducting constructive dialogue and negotiations about an agreement that would supplement the 1951 defense agreement.[61]

9 The May 2004 Response

The official response of the Danish state to the U.S. request was submitted to the parliament in Copenhagen on 26 May 2004 by the then Danish foreign minister Stig Møller.[62] The response, regarded as a positive answer to the December 2002 request by the United States government, indicated that the negotiations between Copenhagen and Nuuk had been conducted based on the principle of mutual benefit.[63] It became clear that the future agreement (that later came to be known as the Igaliku Agreement) would not be restricted to the supplementation or amendment of the old defense agreement (i.e., to the security sphere), but would be a complex agreement that would also include clauses on economic and technical cooperation and cooperation in the field of environmental protection. Especially, the mutual understanding on the economic and technical cooperation between the U.S. and Greenland, which included the establishment of collaboration in scientific research, technology, energy, education, development, tourism, traffic planning and trade, was seen "as a sort of backdoor compensation" by the U.S. for the usage of the

[60] Interview with Adam Worm on 31 July 2017 at the representative office of the Greenland home-rule government in Copenhagen.

[61] Forespørgsel nr.F29. Op. cit. Kl.16:00.

[62] Forespørgsel nr.F60. Op. cit.

[63] Ibid. Kl.11.00.

Thule base (Dragsdahl 2005b). A joint committee by all three sides was set up as a medium for defining the mutual benefit and was expected to exchange information both in formal and informal settings and conduct necessary adjustments by holding regular talks on issues concerning economy and technology.[64] As I shall discuss later, in light of the high expectations for the committee created at the moment of signing of the agreement, its subsequent failure to function properly was conspicuous since it achieved no results.[65] Nonetheless, these expectations regarding the joint committee were clearly stated in the first comprehensive Arctic policy by the Danish state, released publicly in August 2011, which indicates to what extent Denmark and Greenland were keenly interested in the potential for economic development (Udenrigsministeriet et al. 2011: 55).

The Danish government's response most of all meant the consent to the main point of the 2002 U.S. request, i.e., the upgrading of the radar installations in Thule Air Base. In relation to that in the Danish parliament it was emphasized that Greenland had become an equal partner of Denmark and that it had managed to achieve a clear contributory influence on an international agreement. The consent, it was said, meant that in the time to come Greenland and Copenhagen would act in coordination and jointly bear the responsibility for the matter. One of the conclusions was that the response would solidify the standing of Denmark as a NATO country.

Upon hearing the content of the government's response presented by Stig Møller, Denmark's MP Helge Adam Møller commented that the upcoming complex agreement was to be rejoiced because of the following three points. First, the future agreement was to include collaboration in scientific research, as well as economic and technical cooperation, and would thus in the future confer positive influence on Greenland's society. In the sense that it gave it huge inspiration and stimulus, the agreement was more important to Greenland than the U.S.. Second, by allowing the U.S. to use the Thule base as a missile defense outpost, security was being reinforced. Third, Denmark, by acting as a partner and facilitator in the negotiations and by bringing benefits to Greenland, had achieved positive results both for itself and the member of the Danish realm.[66] Nonetheless, during the debate in the Danish parliament it was also added that the consent was being given for the usage of the radar facilities for early warning (i.e., for the introduction of new equipment such as computers, software and communication devices), nothing more and nothing less, and that, as such, the consent to the U.S. request did not exceed the realm of a minor change in the present functioning of the radars in the Thule base.[67] It was empha-

[64] Annex I Joint Declaration by The Government of the United States of America and The Government of the Kingdom of Denmark, including the Home Rule Government of Greenland, on Economic and Technical Cooperation. August 6, 2004, Igaliku.

[65] In that sense, it could be said, the cost of refusing the U.S. request regarding the base is gradually diminishing for Greenland.

[66] Forespørgsel nr.F60. Op. cit. Kl.9.55.

[67] Ibid. Kl.10.40.

sized that, for example if the U.S. wanted to install new, more powerful radars, it would have to ask for a new round of negotiations with Denmark and Greenland.[68]

Kuupik Kleist, a delegate in the Danish national parliament elected in Greenland, although stating that he had not changed his negative opinion on the missile defense project itself,[69] also said that the forthcoming complex agreement would open the door for the normalization of relations between Greenland and the U.S..[70] This is because, based on the premise of the maintenance of equal relations between Greenland and Copenhagen and respect for Greenland's involvement, Thule had become a venue that enabled talks not only about the radar site, but also about trade, the environment, research, tourism, etc. Lars-Emil Johansen too expressed satisfaction by saying that, although not all of Greenland's demands had been reflected, considering that Greenland had fallen into serious difficulties in the past due to secret agreements (i.e., the Hansen document)[71] and that the issue of the radar upgrade affected the land and people of Greenland, he felt content that a fundamental change in the 1951 defense agreement was finally coming and that history was now moving into the right direction, based on a mutual consensus.[72] He also stated that for Greenland this was the first time it was facing the risk and the responsibility posed by the U.S. missile defense shield,[73] but that more than that, especially in relation to the economic and technical cooperation, the expectations were high regarding the island's proximity to North America and the potential of access to its market.[74]

10 The 2004 Igaliku Agreement

Based on the above response of the Danish government, a complex agreement, widely known as the Igaliku Agreement, was signed on 6 August 2004 between the United States, Denmark and Greenland. To be more precise, this was a package consisting of an agreement and two joint statements. Namely, following the content of the response, the package consisted of three parts: the "Agreement to Amend and Supplement the 1951 Agreement on the Defense of Greenland", the "Joint Declaration on Economic and Technical Cooperation" and the "Joint Declaration on Environment".[75] The Igaliku Agreement was structured in such a way due to Greenland's wishes and the understanding that such a composition would prove to

[68] Ibid. Kl.9.15.

[69] Ibid. Kl.10.55.

[70] Ibid. Kl.10.35.

[71] Ibid. Kl.10.40.

[72] Ibid. Kl.10.35.

[73] Ibid. Kl.10.40.

[74] Ibid. Kl.11.00.

[75] Similar to the Itilleq Declaration, as the place of signing the agreement, Igaliku village, the birthplace of Greenland's politician Josef Motzfeldt, one of the signees, was chosen.

be a positive factor in the future. In Greenland, at least, it was believed that even if the military interest in Greenland should fall in the future and the meaning of the defense agreement should come under scrutiny, the arrangements regarding economy, technology and environment could still continue to exist as a separate framework, and that Greenland's relationship with the U.S. would, thus, no longer be defined only by military affairs.

On the other hand, there was concern regarding the functioning of the missile defense, which also had to do with the fact that the agreement was formulated as a reaction to the Danish government's response. For example, during the agreement signing ceremony doubts were expressed about the somewhat naive but crucial question, previously debated in the parliaments, about the potential future development of the missile defense system and the functioning of the missile defense itself.[76] In other words, this is the question of whether the agreement allowed for the possibility of further upgrades of the Thule base in case the U.S. chose to increase its interception capability in order to improve the functioning of the missile defense shield. The then U.S. Secretary of State Colin Powell stated that what was in the Igaliku Agreement was all there was, and that his country had no plans for anything that the local government and Denmark's central government were not aware of. That is, the only thing the U.S. could do (without a new round of negotiations) was to make modest upgrades of software and related functions and equipment, so there would hardly be any changes in the role of the base.

In fact, that there would be hardly any changes in the role of the base is synonymous with saying that no changes would be made in the wording of the 1951 defense agreement, which regulates the existence of the base, and this in itself is fraught with a problem that should not be overlooked. Namely, in the text regarding defense zones in Article I of the 2004 Agreement to Amend and Supplement the 1951 Agreement on the Defense of Greenland, it is stated that Thule Air Base is the only defense area in Greenland. However, Article II of the 1951 defense agreement, which refers to the establishment of defense zones, stipulates that the government of the United States possesses the right "to construct, install, maintain and operate facilities and equipment" in Thule Air Base (Article II, 3b, ii). Furthermore, the article of the amended defense agreement that refers to the NATO Status of Forces Agreement (SOFA) states that "The parties agree that the terms of Article VII of the Defense Agreement shall apply as a supplementary agreement to the NATO SOFA between the parties", but Article VII contains the following text: "All materials, equipment, and supplies required in connection with operations under this Agreement...shall be permitted entry into Greenland free of inspection, customs duties, excise taxes or other charges..." Thus, it seems that the amended version of the 1951 defense agreement does not sufficiently fulfil the conditions for being regarded as the amendment Greenland sought, that is, to be regarded as a relativization of the ruler-and-the ruled relationship Greenland was in with the U.S. in the Cold War era.

---

[76] U.S. Department of State. *Remarks at the Signing Ceremony with Danish Foreign Minister Per Stig Moeller and Greenland Deputy Premier Josef Motzfeldt*. Igaliku, Greenland. August 6, 2004.

11 Conclusion

So, what can we say if we are to summarize this chapter using several concepts defined in Chapter "Base Politics and the Hold-up Problem"? First, needless to say, the structure of the relationship in connection to the Thule base has evolved after the end of the Cold War from a closed relationship between the U.S. and Denmark to a trilateral relationship including Greenland. Greenland has started viewing Thule as an object that it can deal with, and has come to see in the negotiations regarding the inclusion of Thule Air Base into the U.S. missile defense shield a possibility to expand its autonomy (Petersen 2006). It is clear that Thule has value as an effective political bargaining chip not only to Copenhagen, but especially to the post-Cold War Greenland - there is no need to further emphasize how critically important it has been for Greenland.

Furthermore, as has been pointed out in Chapter "Base Politics and the Hold-up Problem", the bargaining regarding military bases was restricted only to Thule Air Base in the 2004 Igaliku Agreement (or to be more precise, in Article I of the Agreement to Amend and Supplement the 1951 Agreement on the Defense of Greenland, which refers to defense zones). Since the possibility of the transfer to mainland Denmark is practically non-existent, the bases are concentrated in Greenland. That is, the substitutability of the Thule base is extremely low and the only partner for bargaining is Greenland, which means that the degree of distribution is also low, which is why, on the whole, the vulnerability of Copenhagen in relation to Greenland is high. Also, the high specificity of the Thule base, located on the 76th Northern parallel in a very harsh Arctic natural environment, has been emphasized in the past, as its main tasks are related to military and security matters. However, it should be noted that its universality (and in that sense low specificity) has been demonstrated empirically, because the base has also been used for civilian purposes as a transportation hub for the locals living in the area, and because, as the Igaliku Agreement shows, broader cooperative relationships not limited to military affairs have been established through it.[77] As it will be argued later, the purpose of the Thule base has been diversifying in parallel with the political convulsions in the Arctic area in recent year - the base is now expected to serve as a platform for international cooperation, so its specificity has been gradually falling.

As already mentioned, the vulnerability of Copenhagen towards Greenland is high, and this basically means that the degree to which Copenhagen may "be held-up" by Greenland is high. For that reason, it was necessary for Copenhagen to design a certain bargaining mechanism regarding the base which would limit instability to a certain scope, i.e., reduce the room for opportunistic behavior. The integration of Greenland, opted for after the Second World War in the period between the expansion of the Thule base and the beginning of its operation, and granting autonomy as a set of rules for incorporating the local voice in a certain framework were the results of such a need. From a theoretical perspective, it can be said that

[77] However, opening the base for civilian use only is not realistic because the maintenance cost (for example the cost of runway maintenance) in that case would be too high.

Copenhagen integrated Greenland in order to reduce its vulnerability in regard to the military base, that is, to reduce the "cost paid for effective adaptation to the changing environment". We have shown in Chapter "Base Politics and the Hold-up Problem" that integration (or incorporation) into own country can be a policy for the nullification of the voice of a subnational actor. As we have briefly mentioned in the introductory chapter, in the theory of local autonomy this is called "functional localization". This is a concept which denotes the treatment of subnational entities that are absorbed by a one-sided rule of the state. This notion has certain explanatory power when it comes to elucidating the actions of central governments, which try to govern their states while keeping down the political cost by enveloping sub-state actors in their legal framework. This phenomenon is conspicuous in the case of the Thule base.

As we shall note in the concluding chapter of the volume, the view that the inclusion of Thule Air Base was an incentive for Denmark to integrate (or, functionally localize) Greenland today seems increasingly convincing. For example, in the first Danish "Strategy for the Arctic", formulated in 2011, two main policy goals are indicated. One is the expansion of Greenland's right to express its views, its autonomy. The other is securing the status of one of the main actors in the Arctic for the Danish state (Offerdal 2014). The Thule base is referred to as a venue that connects these two policy goals, as a hub for future Arctic cooperation, a platform for collaboration. Thule is expected to be used not just for military, but a broader array of purposes, and the understanding that the base indeed has the potential for such usage has been expressed (Udenrigsministeriet et al. 2011: 53). The potential lies in the sea port that can handle the harsh winter conditions of northwest Greenland, tanks and oil storages, factories, hospital, accommodation facilities and other infrastructural and logistic support. Of course, strictly speaking, these expectations are held not only for Thule as an isolated object, but for its coordination with the Sirius Patrol (the dog-sled unit) deployed by the Danish Royal Navy in northeast Greenland and the Joint Arctic Command, of which it is a part (Ozawa et al 2016). Greenland's security depends on the combination of multiple agents and networks, so focusing solely on Thule Air Base would mean ignoring the balance of the military organization as a whole (Namminersornerullutik Oqartussat et al. 2008: 12).

In recent years, though, due to environmental changes in Greenland, such as the melting of the ice sheet, the possibility of exploiting natural resources that have not been accessible in the past and the potential of using Arctic Ocean sea lanes for commercial purposes are growing, which is why it is believed that activities in the Arctic area, including Greenland, will inevitably intensify in the years to come (Lulu 2017; Leira et al. 2016). Thule Air Base is envisioned to function as a hub for various facilities and units. From this point of view too, it possible to see how the substitutability of the Thule base is low (or how its urgency is high). In that sense, the Igaliku Agreement, as a complex set of agreements that guarantee a trilateral relationship involving the U.S., will in the future become increasingly important especially for Greenland as a framework that allows the enhancement of the extent of hold-up. Such a framework that allows it to have its voice reflected in the multifaceted use of Thule Air Base, which is recognized as being of high urgency, can be meaningful to Greenland also in terms of justifying its "threats" as a sub-state actor to the central government. For example, the joint committee established under the Igaliku Agreement as a medium for the fomentation of mutual interests has attracted huge expectations from the Greenland authorities with its wide-ranging scope that includes everything from cooperation in scientific research, technology and energy, to education, development and tourism.

However, the joint declarations on economic and technical cooperation and on environment that make Igaliku Agreement the complex accord it is, in comparison with the amendment of the 1951 defense agreement, have not yielded results that equal the expectations. This is because those two frameworks have been almost ignored since the signing of the Igaliku agreement. The issue of the Thule Base Maintenance Contract (Thule BMC),[78] in which a deal was reached by bypassing Greenland, has been often cited as a clear example of that. The essence of the matter is, however, that the more the high-level urgency and low substitutability of the base are recognized, the more the gap between Greenland's intention to build multifaceted cooperation while enhancing own autonomy by utilizing the Igaliku Agreement and the scale of the issues it is really capable of handling is increasing. In other words, this means that the perspective that captures the bargaining between the central government and the sub-state actor about the military base as a political asset that involves vulnerability and hold-up is, in the circumstances of the conflict between the stable operation of Thule Air Base and the increase of the autonomy of the sub-state actor of Greenland, increasingly becoming a solid ground for a highly effective argument.

References

Ackrén, M. et al. 2015. Greenland as a self-governing sub-national territory in international relations: Past, current and future perspectives. *Polar Record* 51 (259): 407.

Archer, C. 2003. Greenland, US base and missile defense: New two-level negotiations? *Cooperation and Conflict: Journal of the Nordic International Studies Association* 38 (2): 125–147.

Berlingske Tidende. 1949. I Dag underskriver Danmark den nordatlantiske Pagt. April 4: 1.

———. 2001. Dyrt at nægte USA adgang til Thule-radaren. May 29.

———. 2004. Den danske sten i Powells sko. April 30.

Broholt, M. n.d. Den danske debat om Missilforsvaret og Thule-basen. *IIS Research-brief*. No.21: 1–10.

Buchwald, G. et al. n.d. *Atlantpagten: Lande, Folk og Styrker*, 210–214. Odense: Forlaget Codan.

Christensen, S.A. et al. 2009. Greenlanders displaced by the cold war: Relocation and compensation. In *Historical justice in international perspective: How societies are trying to right the wrongs of the past*, 111–134. Cambridge: Cambridge University Press.

Dragsdahl, J. 2003. Grønlandsk missildans: Drømmen om selvstyre skal gennemføres med USAs hjælp. *Information April* 10.

———. 2005a. Denmark and Greenland: American defences and domestic agendas. *Contemporary Security Policy* 26 (3): 486–504.

[78] The contract was transferred from Greenland Contractors, a firm propped up by Greenlandic capital that was up till then in a contractual relationship with the base, to Exelis Services A/S, a company from mainland Denmark, which operates under the umbrella of the U.S. corporation Vectrus, before the negotiations with the Greenlandic firm even ended. 2015. July 7.

———. 2005b. A few dilemmas bypassed in Denmark and Greeland. *Article for Peace Research Institute Frankfurt project on Democratic Peace*.

DUPI: Dansk Udenrigspolitisk Institut. 1997. *Grønland under den kolde krig: Dansk og amerikansk sikkerhedspolitik 1945–68*.

Environmental Law Alliance Worldwide. 2010. *Denmark – HINGITAQ 53 vs. Denmark, Application No.18584/04 (Decision from the European Court of Human Rights as to Admissibility of Application)*.

Gad, U.P. 2017a. Pituffik i Praksis: Nationale reskaleringer i Avanersuaq. *Groenland* 65 (2): 149–167.

———. 2017b. Sex, løgn og landingsbaner: Thule set relationelt. *Tidsskriftet Grønland* 3/2017: 216–241.

Information. 2003. Forurenede aftaler. May 15.

Iwasaki, M. 1998. *Bunken to Renpousei [Decentralization and the federative system]*. Tokyo: Gyosei.

Kawana, S. 2012. *Kichi no Seijigaku: Sengo-Beikoku no Kaigaikichikakudaiseisaku no Kigen [Base politics: The origins of the United States' post-war overseas bases expansion policy]*. Tokyo: Hakuto Shobo.

Kristensen, K.S. 2004. Greenland, Denmark and the debate on missile defense: A window of opportunity for increased autonomy. *DIIS working paper*. Vol. 2004/14: 9.

Kubota, S. et al. 2009. *Huhensei to Sai womeguru porithikkusu: Senjyuumin no Jinruigakutekikenkyuu. "Senjyuumin" toha dareka? [Politics of universality and distinctiveness: An anthropological study of indigenous people. Who are "indigenous people"?]*. Tokyo: Sekai Shisosha.

Leira, H. et al. 2016. Russia and China in Greenland? *NUPI Policy Brief*. No. 43.

Lulu, J. 2017. China, Greenland and competition for the Arctic. *China Policy Institute: Analysis (The online Journal of the China Policy Institute)*.

Lynge, A. 2002. *The right to return: Fifty years of struggle by relocated Inughuit in Greenland. (Complete with an English translation of Denmark's Eastern high court ruling)*. Nuuk: Atuagkat.

Miller, K.E. 1968. *Government and politics in Denmark*. Boston: Houghton Mifflin Company.

Namminersornerullutik Oqartussat. et al. 2008. *Arktis i en brydningstid: Forslag til strategi for aktiviteter i det arktiske område*.

Offerdal, K. 2014. *Interstate relations: The complexities of arctic politics. Geopolitics and security in the arctic: Regional dynamics in a global world*. Abingdon: Routledge.

Olsen, M.R. 2017. Lightning rod: US, greenlandic and Danish relations in the shadow of postcolonial reputations. In *Greenland and the international politics of a changing arctic: Postcolonial Paradiplomacy between high and low politics*. London: Routledge.

Ozawa, M. et al. 2016. *Aisurando, guriinrando, hokkyoku wo shirutameno 65 shou [Sixty-five chapters for knowing Iceland, Greenland and the Arctic]*. Tokyo: Akashi Shoten.

Petersen, N. 2006. *Dansk Grønlandspolitik med særligt henblik på den amerikanske militære tilstedeværelse: Fra stille tosomhed til åbent trekantforhold. Grønlandsk kultur- og samfundsforskning 2004/05*. Nuuk: Ilisimatusarfik/Atuakkiorfik.

Taagholt, J. et al. 2001. *Greenland: Security perspectives*. Alaska: ARCUS.

Tamnes, R. et al. 2014. The geopolitics of the arctic in historical perspective. In *Geopolitics and security in the arctic: Regional dynamics in a global world*. London/New York: Routledge.

Ueda, T. 2007. Reisensyuuketsugo no Beikoku to Taiseiyou-doumei [The Unites States and the Atlantic Alliance after the end of the cold war]. *International Politics*. 150: 100.

Udenrigsministeriet. et al. 2011. *Danmark, Grønland & Færøerne: Kongeriget Danmarks Strategi for Arktis 2011–2020*.

United States Strategic Air Command. 1979. History of the strategic air command, 1 January 1958–30 June 1958. *Historical Study*. No.73: 88–90.

Yoshitake, N. 1985. Denmaaku no Doumeigaikou: NATO Kameisyoki (1949-54nen) no Kichimondai wo cyuushintoshite [Denmark's alliance diplomacy: Focusing on the issue of military bases in the early years after joining NATO (1949–54)]. *Research on Nordic History* s4.

How Have the U.S. Interests in Greenland Changed?: Reconstructing the Perceived Value of Thule Air Base After the Cold War

Kousuke Saitou

Abstract The purpose of this chapter is to examine how the U.S. interests regarding Thule Air Base have changed in the post-Cold War period. Initially, various U.S. bases in Greenland were established or strengthened to counter the Soviet threat in the early stages of the Cold War. Therefore, when that threat receded, it was expected that their strategic importance would diminish dramatically. However, by the end of the 1990s, the U.S. recognized the increased missile capabilities of "rogue states," accelerating the deployment of the missile defense system. Thus, the value of Thule Air Base increased as a key part of the system, affecting the trilateral negotiations between Washington, Copenhagen, and Nuuk. The U.S. military bases in Greenland also serve as hubs for scientific research concerning the Arctic region, although this feature affects base politics less directly than do military factors, due to third-party influences, such as civilian infrastructure and international cooperation.

Keywords Anti-ballistic missile (ABM) treaty · Arctic Research Commission · Ballistic missile early warning radar (BMEWS) · Missile defense · National Science Foundation (NSF) · Rumsfeld Commission Report · Scientific research · Sondrestrom Air Base · Strategic Defense Initiative (SDI) · Thule Air Base

1 Introduction

Using Thule Air Base as a bargaining chip, Greenland has conducted negotiations with Copenhagen and Washington, striving to expand its autonomy. If the strategy works well, a premise would be established: that the United States, the installer of the base and the "superior actor" in the trilateral negotiation framework, has continued to recognize a certain value in Greenland or, more specifically, Thule Air Base.

If we look at the current situation, it would seem that the value of Greenland for the U.S. can be explained as follows. First, the radar site at Thule Air Base is an

K. Saitou (✉)
Research Initiatives and Promotion Organization, Yokohama National University, Kanagawa, Japan
e-mail: saitou-kousuke-ps@ynu.ac.jp

M. Takahashi (ed.), *The Influence of Sub-state Actors on National Security*, Springer Polar Sciences, https://doi.org/10.1007/978-3-030-01677-7_4

important part of the U.S. missile defense system that is designed to prevent a missile attack on the U.S. homeland, although the deployment scale is still limited. Second, due to its distinctive geographic feature, the fact that it is "the base closest to the Arctic," the strategic value of Thule Air Base is increasing because much attention is focused on the Arctic region. Third, Greenland itself serves as a hub of scientific research in the Arctic region, although this is not receiving so much attention in the field of security studies or international relations. Thule Air Base gives partial but important support for these activities in the harsh environment of Greenland, where civilian infrastructure is minimal.

However, why has Thule Air Base been maintained even after the end of the Cold War? The answer is not directly derived by reference to the above-mentioned present significance of Thule Air Base. We must examine, from the viewpoint of the history of the U.S. base policy, how the value of Thule Air Base has been changing since the end of the Cold War, which accelerated global closure of U.S. bases. The bases in Greenland were established and strengthened to counter the Soviet threat in the early stage of the Cold War (Petersen 2011), so their strategic importance dramatically diminished when the threat from the Soviet Union receded. The threat reduction was coupled with a severe fiscal pressure on the U.S. defense budget in conjunction with the "twin deficits" problem that occurred in the 1980s. In this military and fiscal climate, the closure of Thule Air Base, which requires relatively high operating costs, was supposed to be considered.

In reality, the base stayed. One explanation for this is that Greenland is geopolitically important to the U.S. missile defense system, which is central to the U.S. national security strategy today. Nonetheless, in the context of the missile defense study, another point of interest remains. That is, there was much skepticism about the system during the 1990s due to the cost and issues of technical feasibility, which is why its deployment was not so loudly approved as it is today. The discourse justifying its development and deployment has been gradually constructed over time in the U.S. political process, raising various technical and strategic concerns (Peoples 2010).

Even if we focus solely on the context of scientific research in the Arctic region, there is still room to consider how the importance of the base was perceived in the past and how it has changed. Although the value of Thule Air Base has been reassessed in recent years because of the rising importance of Arctic issues, it was not until the 2000s that world's interest in the Arctic region significantly increased. In relation to this, it is also necessary to discuss how the military bases in Greenland exhibited their value as hubs for scientific research in the Arctic region. In fact, the military played a major role in the development of scientific research in the Arctic after the Second World War. It has been pointed out that the U.S. military became a patron of research in the field of the earth sciences because of its military own needs, but that in the same way, the scientific community too may have required needed this support by the military for its activities (Doel 2003). Given the major importance of Greenland for scientific research, it is also worth reconsidering how this aspect of Greenlandic value was perceived by the U.S. in the process of the base realignment.

This chapter addresses why and how the U.S., a "superior actor" in the trilateral relationship, has changed its perception on Greenland and Thule Air Base after the end of the Cold War, within the context of the drastically transforming international security environment. This question overlaps with a more general discussion of base politics: why has the U.S. maintained some overseas bases even after the original conditions and strategy which required their installation changed? In the sections below, the main task is to shed light on the process of the construction of the U.S. perception chronologically, rather than derive the logic of the continuity of the base from its current strategic significance. In addition, it is also worth turning our attention briefly to Sondrestrom Air Base, which was also located in Greenland, but closed in the early 1990s, as a reference point. These two bases in Greenland, Thule Air Base and Sondrestrom Air Base, were treated in different ways under the drastically changing security conditions. As we shall see later, it is no wonder that Sondrestrom Air Base was closed at the end of the Cold War. However, parallel with the calls for closure, there were also some calls within the U.S. to maintain it. The argument by those supporting Sondrestrom Air Base is also indicative of the value of the bases in Greenland.

2 Military Value of Thule Air Base and its Cost

2.1 The Cold War and Greenland

U.S. interests in Greenland have been, first and foremost, based on its military-strategic requirements. During the Second World War, the awareness of the value of Greenland, which became a foothold for approaching the European theater, greatly rose in the U.S. On the basis of the Defense of Greenland Agreement signed with Denmark in 1941, many military bases were installed on that Arctic island. A list made by the Joint Chiefs of Staff in 1946 shows six "essential" overseas bases, "three of which (Greenland, Iceland, and the Azores) were declared to be of 'outstanding importance'."[1] Even after the end of the Second World War, both the U.S. and European countries have paid more attention to Greenland, because of the arrival of the Cold War, during which the technological development of missiles and long-range bombers progressed and came to critically influence the U.S. national security. As a result, the 1941 Agreement was renewed in 1951, leading to the construction of Thule Air Base. It was expected to function as a base for B-36 bombers, while reconnaissance aircraft, aerial-refueling aircraft, and, subsequently, nuclear weapons, were also deployed there. Moreover, together with the Defense Early Warning (DEW) radar, which was intended to give early warnings about bombers flying from the Soviet Union, the United States introduced to Greenland the Ballistic Missile Early Warning System (BMEWS) for detecting ICBMs coming out of the

[1] Petersen, N. 2011. "Greenland in the U.S. Polar Strategy": 92.

Soviet Union. This formed an important part of the military value of Thule Air Base up to the present day.

This brief postwar history suggests that, from the military point of view, the value of Greenland for the U.S. was coming from the context of its counter-Soviet strategy. Therefore, its strategic importance was expected to be greatly shaken by the end of the Cold War. As Archer pointed out, one of the major differences between the post-Cold War and previous period is whether a serious threat like the Soviet bloc existed or not, which affected the attitude of Denmark and Greenland in their negotiations (Archer 2003). Such a change in threat had, naturally, the same consequences for the U.S. security strategy and its diplomatic attitude. Furthermore, during the 1980s, the U.S. was facing the so-called problem of "twin deficits" that imposed a severe pressure on its national defense budget, on which military base management naturally depends. These two factors led to the movement for base realignment and closure from the end of the 1980s onwards. In other words, as the military threat declined, and the policy goal of reducing the fiscal deficit rose in importance, there was increasing motivation to close overseas bases. How did the bases in Greenland fare?

2.2 Threat Reduction, Fiscal Pressure, and the Value of Thule Air Base

A survey report released just after the end of the Cold War shows that Thule Air Base, along with Naval Air Station Keflavík in Iceland, was one of the largest American bases in terms of the investment in facilities. As of 1992, the U.S. invested a total of $7 billion in its bases in Europe. Of that sum, the largest amount, $450 million, was invested in Keflavík, and the next largest was Thule Air Base at $377.7 million, followed by Stuttgart, the base with the largest investment in continental Europe, at $255.7 million,[2] although it is worth noting that these bases had different functions.

The U.S. Congress argued intensely over the base-closure issue from the late 1980s on. However, the U.S. Department of Defense planned to invest in a renewal of the facilities at Thule Air Base, together with the facilities of several other bases in the European and Pacific regions. Of course, it was questioned at that time whether this investment plan was "an aberration" from the trend toward less overseas military construction. James Boatright, Deputy Assistant Secretary of the Air Force for Installation, explained that the investment in base construction had still been significantly reduced compared to the previous request and was not only

[2] Grimmett, R. F. 1992. *Base Closures in Europe: Cost and Procedural Issues*, CRS Report for Congress, July 27: 15–20. By country, the investment amount of Germany is the largest, at approximately $3.891 million, followed by the United Kingdom in an extremely distant second place ($794.7 million).

important for supporting military operations, but also critical to maintaining good relations with the host nations.[3]

In his statement submitted to the Armed Services Committee hearing, Boatright reiterated that the plan of the Department of Defense was in compliance "with the guidance by this committee and the Congress to reduce our overseas MILCON (military construction) expenditure to the lowest possible level," including the budget for constructing military bases in Greenland, and that the DoD was looking to, if possible, "shift to host nations the financial burden of supporting U.S. military personnel stationed overseas."[4] As part of this trend, the construction investment in Greenland was justified from the viewpoint of operational cost efficiency. The projects at Thule Air Base, related to the construction of dormitories and a runway, would reduce the extremely high maintenance and energy costs of demolition of the old dormitories. According to an estimate of the Department of Defense, the projects would "pay back their investment costs in less than 5 years."[5]

The investment in Thule Air Base, like many other overseas bases, was certainly affected by severe fiscal pressure and was checked against the criterion whether it could make its operations more efficient or not. However, the military value of Thule Air Base itself was changing. In fact, aside from the argument linking the threat reduction to the base closure, some emphasized the different functions of Thule Air Base from those during the Cold War and attempted to justify the continuation of the base. One of these claimed that Thule Air Base played a role in fulfilling the U.S. space-related mission requirement. The Gulf War demonstrated that the capabilities of the new U.S. military were supported by the development of ICT dependent on the space infrastructure, such as GPS. The experience helped the U.S. become aware of the increasing importance of space, or of the space-related military assets, even after the end of the Cold War.[6] FY1994 national defense request included a series of space-related military construction programs to be implemented in overseas bases, like the investment in the space launch infrastructure project at Antigua Air Station, a satellite-tracking facility at Diego Garcia, a GPS satellite control facility, and wastewater treatment plants at Thule Air Base. "These projects all support U.S. space programs" and "are not affected by any force structure reduction in Europe," the Air Force stated.[7]

---

[3] Patricia Schroeder (Chairman) and James F. Boatright (Deputy Assistant Secretary of the Air Force, Installations), Hearing before Military Installations and Facilities Subcommittee, the Committee on Armed Services, Senate, March 19, 1992: 193.

[4] James F. Boatright, Hearing before Military Construction Appropriations Subcommittee, the Committee on Appropriations, House of Representatives, March 12, 1992: 484.

[5] Ibid.: 484–485, 531.

[6] Posen pointed out the military "Command of Commons" in air, sea, land, and space as a source of the U.S. hegemony in the post-Cold War era, as well as its limitations. Posen, B. R. 2003. "Command of the Commons: The Military Foundation of U.S. Hegemony," *International Security*, 28, no. 1: 5–46.

[7] Gary D. Vest (Deputy Assistant Secretary, Air Force) and James E. McCarthy (civil engineer, Air Force), Hearing before Military Installations and Facilities Subcommittee, the Committee on Armed Services, House of Representatives, April 22, 1993: 17.

2.3 Missile Defense, ABM Treaty, and Thule Air Base in the Post–Cold War Period

The U.S. was forced to consider the legal consistency of the Strategic Defense Initiative (SDI) that was planned by the Reagan administration, and of the construction of the missile defense system that was subsequently implemented, with the Anti-Ballistic Missile (ABM) Treaty. In 1972, the U.S. and the Soviet Union had signed the ABM Treaty, by which they vowed not to deploy ABM systems except for their respective national capitals and an area containing ICBM silo launchers. The ABM Treaty was updated in 1974, to further limit the deployment area to one. This treaty was based on the shared U.S. and Soviet recognition of the deterrence effects, the so-called "mutually assured destruction (MAD)". Under MAD, a nuclear stalemate is intentionally established because use of nuclear weapons would lead to the mutual destruction of both sides. In addition, the treaty was also intended to respond to concerns that the effectiveness of missiles would be diminished by the technology to intercept them and that, as a result, the nuclear arms race would intensify.

One of the concerns about introducing a missile defense system was its legal consistency with the ABM Treaty, from which the U.S. decided to withdraw later due to its accelerating missile defense program. The deployment of a radar system at Thule Air Base proved no exception. The Soviet Union pointed out the possibility that the radar system in Greenland was in violation of the treaty immediately after it was signed. In order to address these concerns, the Soviet Union requested permission to visit the bases in Greenland, but this request was denied. The use of the radar site in Greenland in connection with ballistic missile defense had been a long-standing point of contention between the U.S. and the Soviet Union, subsequently Russia.[8]

More importantly, the U.S. recognized that this issue had become politicized in Denmark and Greenland, with which it had to negotiate how to handle the status of the radar site there. In 1987, the Danish Parliament refused to use the radar at Thule Air Base as a part of the SDI. However, in response to this, the Danish government issued an official opinion in 1989 that the deployment of the radar was not contrary to the spirit of the ABM Treaty (Fischer 1993). At any rate, there was certainly division within Denmark between those supporting and opposing the Danish involvement in the SDI project.

Because of the legislation of the Danish Parliament, prohibiting the use of Thule BMEWS radar in connection with SDI or an ABM system, one of the discussion points in the U.S. Congressional committee hearings in the early post-Cold War era was whether it would be permitted to use the radar in Thule Air Base in such a

---

[8] U.S. Department of State, Soviet Statement in Connection with the Third Review of the Treaty between the United States of America and the Union of Soviet Socialist Republics on the Limitation of Anti-Ballistic Missile Systems, September 1, 1988, https://www.state.gov/t/avc/trty/101888.htm (accessed March 20, 2018).

system or not. The Department of Defense told the Danish government that upgrading the BMEWS network as a part of the missile defense system was one option, which had been examined but not decided yet. Stephen Hadley, Assistant Secretary of Defense, explained that any decision that "would involve facilities on Danish territory would be made only after coordination with the Danish government."[9] The U.S. government seemed to be trying to resolve this conflict through negotiations, rather than ignoring it and simply proceed with the deployment anyway. As described above, the plan to give Thule Air Base a missile defense function was, at least officially, no more than one option that had not necessarily been determined. However, the U.S. simultaneously recognized the importance of Thule Air Base itself, being aware that the negotiation with the Danish government would be indispensable for using it. This attitude of the U.S. would subsequently lead to three-party negotiations on the missile defense system in Greenland.

3 Greenland as a Hub of Scientific Research: The Withdrawal from Sondrestrom and the Value of Thule

3.1 The Withdrawal from Sondrestrom Air Base and its Influence on Scientific Activities

The U.S. interests in the Arctic region from a scientific perspective are as old as its military interests there. The National Academy of Sciences established the Committee on Polar Research in 1958 in response to a request from the Director of the National Science Foundation (NSF). Since then, it has conducted research activities in the Arctic and Antarctic regions. In 1983, the basic components of the U.S. Arctic policy were confirmed by the National Security Decision Directive issued by President Ronald Reagan. It included the protection of essential security interests, support for sound and rational development while minimizing adverse effects on the environment, promotion of scientific research, and promotion of mutually beneficial international cooperation in the Arctic to achieve the above objectives (Reagan 1983). The Arctic Research and Policy Act was enacted in 1984, establishing the U.S. Arctic Research Commission to promote Arctic research and advise on the policy. The U.S. involvement in scientific research in the Arctic region was steadily institutionalized through this legislation including a designation of "the National Science Foundation as the lead agency responsible for implementing Arctic research policy."[10]

A series of debates on the withdrawal from Sondrestrom Air Base, together with the growing U.S. interest in the Arctic, helped clarify the importance of the hubs for

---

[9] Samuel Nunn (Chairman) and Stephen J. Hadley (Assistant Secretary of Defense for International Security Policy), Hearing before the Committee on Armed Services, Senate, May 20, 1992: 537.

[10] P.L. 98–373, Arctic Research and Policy Act of 1984, July 31, 1984.

scientific research in Greenland and the role of military bases for that objective. The U.S., in the post-Cold War period, had sequentially realigned its military bases worldwide. Sondrestrom Air Base was included in a list of bases from which the U.S. planned to withdraw "without being formally requested" by the host nations."[11] According to a document submitted in a Congressional hearing, the U.S. had conducted negotiations with the Danish government on the withdrawal from Sondrestrom Air Base in 1990, and it was returned to the Greenland Home Rule Government on September 30, 1992, while the U.S. maintained access rights.[12]

It was not the military, but the U.S. scientific community, that questioned this withdrawal plan. During the hearing session entitled "Arctic Ocean Research," various concerns and requests of the NSF were presented as the views of the Arctic Research Commission. First, if Sondrestrom Air Base was returned to Denmark, the Department of Defense was to consider maintaining appropriate logistical support capability for the scientific research programs that were important to the U.S. Second, the relevant government agencies, including the NSF, were to immediately alert the U.S. scientific community involved in the research in Greenland and provide the details of the ongoing negotiation on Sondrestrom Air Base. Third, several alternative scenarios on returning the base were to be considered, and the corresponding financial impacts for each existing or planned research project were to be estimated. Fourth, a new approach for research activities in Greenland was to be designed in a more cost-effective manner. Finally, closer research coordination with Danish and other European researchers was to be developed, so that all parties shared equitably the increased cost of logistics.[13] In sum, the withdrawal from Sondrestrom Air Base was understood to be a comparatively important issue, even for the scientific community engaged in Arctic research.

Senator Al Gore, in a subcommittee on "Monitoring the Arctic and Antarctic Environments," asked what impact the return of Sondrestrom Air Base would have. In this subcommittee, an NSF staff member responsible for polar research activities predicted that when the facilities in Sondrestrom Air Base came to be managed by the Greenland Home Rule Government, it would no longer be possible to carry out the activities conducted by piggybacking onto the Air Force flights and facilities, which would especially affect the Ice Core Program. The NSF also pointed out that the costs of the activities were expected to go up due to the loss of Air Force facilities. At that time, the Greenland Home Rule Government informed the U.S.

---

[11] A prepared statement in a hearing on base closures, Hearing before the Military Installations and Facilities Subcommittee, the Committee on Armed Services, House of Representatives, February 27, 1991: 215.

[12] In the 1991 MOU on the withdrawal, it was recognized that on the one hand the non-removable property left at Sondrestrom Air Base was to be relinquished, but that on the other hand, during a crisis the U.S. military would be entitled to priority use of the runway, parking, and refueling facilities.

[13] A prepared statement in a hearing on Arctic Ocean research, Hearing before the National Ocean Policy Study of the Committee on Commerce, Science, and Transportation, Senate, April 24, 1991: 50. See also U.S. Arctic Research Commission, *Logistic Support of United States Research in Greenland: Current Situation and Prospects*, December 1990.

government that it would be required to cover the costs of running the airport and the facilities, to bring in the fuel, and to operate the traffic control system and the meteorological system. Such costs would be obviously different from those that were incurred while relying on the support of the Air Force. Thus, the NSF was concerned about the increase of the research expenditure in Greenland due to the withdrawal.[14]

3.2 Does the Characteristic of Being a Scientific Research Hub Contribute to the Construction of the Value of a Base?

As the importance of scientific research in the Arctic region continually increases, the role of Greenland as a research hub is expanding due to its geographical characteristics. Then, if the withdrawal from Sondrestrom Air Base was unavoidable in the post-Cold War period, where would the U.S. seek a foothold to replace the one it had lost? It is possible or reasonable to presume that its role was taken over by Thule Air Base, which continued to be maintained in Greenland. However, as a matter of fact, the U.S. does not necessarily depend only on Thule Air Base for scientific research, despite being increasingly aware of the value of Greenland as a scientific research hub.

The first point is that Sondrestrom Air Base continued to maintain a part of its functions as a private-sector hub airport, while the management was transferred to the Greenland Home Rule Government. In addition, the NSF and the Greenland Home Rule Government had negotiated its continued use for the purpose of scientific research. Therefore, the withdrawal itself would not cause immediate and serious damage to those kinds of activities (Kelly 1992). Even today, in accordance with the MOU agreed in 1992, the facilities located at Kangerlussuaq can be used for scheduled U.S. military flights and contingency operations. The base has continued to serve as a logistical support base for scientific research, including science camps (that are conducting research on the Greenland Ice Cap) of the U.S. and European countries.[15]

The second point is that the logistical support for scientific research in Greenland has become resilient based on several approaches. In a report on the logistics published by the Arctic Research Commission in 1997, the importance of Thule Air Base was certainly emphasized. This report pointed out the geographical characteristic of Thule, being in northwestern Greenland, and explained that it provided unique opportunities for scientific research on the high-latitude polar region, the

[14] Al Gore, Robert Corell (Assistant Director of Geosciences, NSF), and Peter Wilkniss (Division of Polar Program, NSF), Hearing before the Subcommittee on Science, Technology, and Space, the Committee on Commerce, Science, and Transportation, Senate, May 13, 1991: 46–47.

[15] U.S. Department of Defense, Office of Undersecretary of Defense (Policy), *Report to Congress on Arctic Operations and the Northwest Passage*, 2011: 18.

Arctic Ocean, and other marine environments.[16] However, not all capabilities for scientific research necessarily depend on military facilities The report also indicated that the U.S. had obtained a wide range of logistics in Greenland in recent years, while recognizing that the U.S. military facilities in Thule Air Base should still be utilized for the purpose of scientific research. It included not only Thule Air Base, but also the support from the Danish government, the Greenland Home Rule Government, and other civilian transportation infrastructure.[17]

This dispersive approach seems to assume that Thule Air Base might be placed in a politically unstable position sooner or later. The 1990 Report of the Arctic Research Commission spoke of the risk of Thule Air Base being returned within 10 years like Sondrestrom Air Base.[18] At a committee hearing, when the candidate sites for the Arctic ice cap observation base were narrowed down to Resolute Bay in Canada, Thule Air Base was also listed as an alternative site. At the same time, it was also pointed out as problematic that Thule Air Base was intended for military purposes without civilian infrastructure, so the access to it would be lost if the base was closed.[19] It is not clear whether the factors causing the political instability of Thule Air Base were derived from the severe budget pressure within the U.S. or the political movement opposing the base that was smoldering within Greenland.

Whatever the case, the U.S. scientific research activities in the Arctic region depended to a certain extent on the U.S. military equipment and capabilities. In that sense, the presence of a base in Greenland had important value for the U.S., not only for military, but also for scientific purposes. These purposes and measures were intertwined, constructing the value of Greenland. However, the scientific value of Greenland for the U.S. was secured in several different ways so it is better not to think that the scientific activities were immediately impaired by the withdrawal from Sondrestrom Air Base and that the dependence on Thule Air Base for research dramatically increased.

4 Missile Defense and the Igaliku Agreement

4.1 Influence of the Missile Technology Diffusion

U.S. demands for the deployment of the missile defense system increased rapidly at the end of the 1990s. Initially, the Clinton administration did not actively promote the introduction of the missile defense system (Spring 2000). However, the

---

[16] U.S. Arctic Research Commission, *Logistic Recommendations for an Improved U.S. Arctic Research Capability*, July 1997: 83, https://www.arctic.gov/publications/other/usarc_logistics_report.html (accessed March 20, 2018).

[17] Ibid.: 15.

[18] U.S. Arctic Research Commission, 1990: 10.

[19] Post-hearing questions submitted to Neal Lane (Director, NSF), Hearing before Subcommittee on Basic Research, the Committee on Science, House of Representatives, April 22, 1998: 108.

publication of the so-called "Rumsfeld Commission Report" on the missile threat assessment, on July 15, 1998, increased awareness of the importance of the missile defense system. This report noted that "rogue states," like North Korea and Iran, might be able to acquire ballistic missiles capable of inflicting great damage on the U.S. within 5 years and that Iraq might acquire them within 10 years. The report also expressed concerns about the uncertainty of missile control in Russia, which could cause accidental missile launches. In addition, it mentioned the possibility of China provoking the U.S. over the Taiwan Strait issue and that the transfer of missile technologies to Pakistan and Saudi Arabia could bring about new threats.[20] The findings of this investigation were taken seriously in the U.S. security community, increasing greatly the awareness of the threat posed by the diffusion of ballistic missile technology. It led to strong support for the rapid construction of the missile defense system and accelerated the movement to withdraw from the ABM Treaty.

Under the National Missile Defense Act, enacted in 1999, it was decided "to deploy as soon as is technologically possible an effective National Missile Defense system capable of defending the territory of the United States against limited ballistic missile attack."[21] However, the attitudes of the allies in whose countries the system was intended to be deployed were also very important for the U.S., in terms of political feasibility of its implementation. Around this time, although the first missile interception test was successful, the subsequent tests failed. This led to some skepticism about the feasibility of the missile defense system. The U.S. government had to respond to these concerns. At a public hearing on missile defense in 2000, the armed services committee opined that the progress of technological development must be explained to U.S. citizens and to the world, as a matter of commitment to the missile defense system. William Cohen, Secretary of Defense, commented as follows:

> In order to have a technologically effective system, we need to have the support of our allies. If we do not have the support of our allies with respect to forward deployed X-band radars, you will not have an effective technologically reliable system.[22]

In other words, he stressed again that the U.S. needed to persuade its allies in order to proceed with the deployment of the missile defense system, which was a diplomatic problem. Senator John Warner, the chairman of the committee, confirmed the need for the cooperation of the United Kingdom and Denmark when deploying the radars at Fylingdales and Thule, with the comment that "they are sovereign…nations."[23] Thus, the issue of the technological feasibility of the missile defense system was linked to the necessity of cooperation with the host nations.

[20] Executive Summary of the Report of the Commission to Assess the Ballistic Missile Threat to the United States, July 15, 1998, https://fas.org/irp/threat/bm-threat.htm (accessed March 20, 2018).

[21] P.L. 106–38, National Missile Defense Act of 1999, July 22, 1999.

[22] William Cohen (Secretary of Defense) and John Warner, Hearings before the Committee on Armed Services, Senate, July 25, 2000: 101.

[23] Warner, Hearings before the Committee on Armed Services: 137.

4.2 Dialogue with Divided Greenland

The U.S. effort toward integrating Greenland into the missile defense system was continued within this political climate. A three-party meeting among the U.S., Denmark, and Greenland was held in Nuuk in August 2000. A report says John Holum, the U.S. Under-Secretary of State for arms control and international security, attended the meeting and stated that "they didn't ask for anything concrete. It was just a discussion." However, it was also noted that the Clinton administration was to decide whether to introduce the system within a few weeks, so there seemed to be time pressure.[24]

Regarding the meeting, Jonathan Motzfeldt, then-Prime Minister of the Greenland Home Rule Government, said that the U.S. plan was "neither an offensive nor a nuclear project," and so it could be categorized "within the ambit of the normal use of the U.S. radar base at Thule." Therefore, he did not consider it to be problematic. On the contrary, some members of the Greenlandic Government opposed the deployment of the missile defense system, based on the understanding that it would lead to a new arms race.[25] Similarly, in another interview, Motzfeldt said that "no one in Greenland wishes to take actions that would lead to recreating the atmosphere of the Cold War era," and "I am content that NATO has not greeted the National Missile Defense plans with cheers." (Brooke 2000)

There were also various opinions outside of the government. A newspaper suggested that Greenland's indigenous people were worried by Russia's opposition to Greenland's engagement on the front line of the missile defense system, and there were concerns that their peaceful territory would be transformed into "ground zero." At the same time, Lars Jeremiassen, then-mayor of Qaanaaq, stated that the recent developments presented "a chance to get our old hunting and burial ground back." (Williams 2000)

It is debatable whether this situation was caused by the "divided" political voices in Greenland, or merely by the "non-determined" attitude of the Greenlandic people including the Prime Minister. However, either interpretation supported Greenland using a radar upgrade in Thule Air Base as a bargaining chip for the expansion of its autonomy because the more domestic political situations were divided, the stronger Greenlandic negotiator's power.[26] In addition, the stronger the interest the U.S. showed in maintaining and militarily, diplomatically, and scientifically strengthening Thule Air Base, Greenland would be able to use Thule Air Base as the more powerful bargaining chip in the negotiations. Greenland had a great opportunity to

---

[24] "Greenland Divided over US Missile Shield," *Space Daily*, August 23, 2000, http://www.spacedaily.com/news/bmdo-00zzy.html (accessed March 20, 2018).

[25] Ibid.

[26] This can be clearly inferred from the logic of two-level games. See Putnam, R. D. 1988. "Diplomacy and Domestic Politics: The Logic of Two-Level Games," *International Organization*, 42, no. 3: 427–460. Though from a slightly different viewpoint, Archer also analyzes the history of trilateral negotiations among Copenhagen, Nuuk, and Washington D.C. with the theoretical framework of the two-level games. Archer, "Greenland, US Bases, and Missile Defence": 125–147.

reconsider and to decide its own position in the negotiation because of the value of Thule Air Base being directly linked to the various interests of the U.S.

4.3 Upgrading the Missile Defense Radar in Thule Air Base

In the 2000s, the budget for missile defense was not such a major controversial issue as to overturn the implementation of the project. It indicated that the plan to upgrade the radar in Thule Air Base was already beyond the point of no return. The U.S. was deeply dependent on Greenland for the missile defense deployment. In the FY2002 Congressional hearing on the military construction budget, Senator James Inhofe referred to the airfield repairs at Wake Island ($25 million) and Thule Air Base ($19 million) as the Air Force's two largest construction projects and questioned their necessity after positioning them as "minor bases," which had no significant Air Force mission. The Air force objected, pointing out that Thule Air Base was playing an important role in the warning and surveillance of sea-launched ICBMs, satellite tracking and control, and support for *en route* airlift operations during contingencies. In order to maintain such a capability, it was necessary to improve the airfield and the quality of life of the soldiers stationed there.[27]

In 2003, immediately before the conclusion of the Igaliku Agreement, the negotiations with Denmark and Greenland were mentioned in terms of the investment in the missile defense system in Thule Air Base. At a Senate committee hearing, the Air Force clarified its effort to advance a plan for upgrading the early warning radar of the missile defense system in the UK and Greenland, and to obtain their consent. It was also reconfirmed that Denmark's consent should be obtained for the radar upgrade in Thule Air Base. As a matter of fact, the UK had already agreed to the radar upgrade at Fylingdales (scheduled to be completed by late 2005). It was also revealed that Denmark had indicated its intention to approve the radar upgrade at Thule Air Base (scheduled to be completed by mid-2006), by updating the 1951 Greenland Defense Agreement. After all, there were minor concerns within the U.S. at the time that the progress of the negotiations had a serious effect on the missile defense issue.[28]

It was also notable that the push for realizing the missile defense system accelerated globally during this period. Allies like Australia, Canada, and NATO expressed their intention to participate in the system.[29] This indicated that the U.S. interest in deploying the missile defense system in Greenland was increasing. Since missile defense operates as a "system," the influence of its elements on the U.S. security

---

[27] James M. Inhofe, and Earnest O. Robbins II (USAF, Civil Engineer), Hearing before the Committee on Armed Services, Senate, August 2, 2001: 206–207.

[28] Ronald T. Kadish (USAF, Director, Missile Defense Agency), Hearings before the Committee on Armed Services, Senate, March 18, 2003: 614, 616, 670.

[29] Ronald T. Kadish, Hearing before Subcommittee of Strategic Forces, the Committee on Armed Services, House of Representatives, March 25, 2004: 204.

strategy increased as its scope widened. The so-called "network effect" theory[30] can explain this mechanism: the effects of the missile defense system increased as the number of participating countries rose and they shared information with each other, while the area that the system was able to cover expanded. As a result, the costs of withdrawing from the bases that had these core functions, including Thule Air Base, also rose. In other words, the U.S. dependence on the relevant region would rise, leading to an increase in its value.

5 The Expanding Value to the Present

The negotiation was, for both the U.S. and Greenland, concluded as a mutually beneficial agreement. Colin L. Powell, U.S. Secretary of State, emphasized its significance to accelerating the missile defense project, while Josef Motzfeldt, Deputy Prime Minister of the Greenland Home Rule Government, argued that it was "the day when Greenland took a decisive step toward equality and co-responsibility."[31] Even after the Igaliku Agreement, the U.S. diplomatic considerations toward Greenland seem to be unchanged. The U.S. stance regarding the construction of missile defense components abroad was ultimately to be based on the consent of its allies. Henry Obering, director of the Missile Defense Agency, explained that the requests for the FY2006 defense budget included the project to upgrade the radar at Thule Air Base, which would "play an important role in the system by giving us [the U.S.] an early track on hostile missiles." Obering also stated that negotiations had been repeatedly conducted with the Greenland Home Rule Government, and its consent had been obtained for the implementation of the project.[32] The budget request in the following year also included the Thule Air Base reinforcement. It was justified with the conclusion of the U.S.-Danish agreement, which would make it possible to implement the radar upgrade project at Thule Air Base by 2009.[33]

In Europe, the U.S. continued to negotiate with Poland and the Czech Republic on the deployment of missile defense assets. The successful negotiations with the UK and Denmark likely showed the political feasibility of missile defense both domestically and internationally. In addition, the importance of comprehensively using the systemic elements in California, Alaska, Fylingdales, and Thule in order to cover the whole U.S. territory from missiles launched from the Middle East was

---

[30] Network effects refer to the phenomenon in which, alongside the increase in the number of participants, the value of the network rises and the benefits for the users increase.

[31] The Associated Press, "Greenland Base to be Upgraded as Part of Missile Shield Plan," *New York Times*, August 7, 2004, http://www.nytimes.com/2004/08/07/world/greenland-base-to-be-upgraded-as-part-of-missile-shield-plan.html (accessed March 20, 2013).

[32] Henry A. Obering (USAF, Director, Missile Defense Agency), Hearing before the Committee on Armed Services, Senate, April 7, 2005: 153, 160.

[33] Henry A. Obering, Hearing before Strategic Forces Subcommittee of the Committee on Armed Services, House of Representatives, March 9, 2006: 6–8.

emphasized.[34] As Obering pointed out, "without the participation of our [the U.S.] allies and friends, the ballistic missile defense system would look very different."[35] Eventually, the utility of the whole missile defense system increased, while the importance of each component that undertook the core functions, such as Thule, rose to an unprecedented level.

It was not only the acceleration of the deployment of missile defense components, but also the transformation of the security environment in the Arctic, that indicated deeper U.S. dependence on Thule Air Base. The importance of the Arctic region was, for the first time, described in the National Security Strategy 2010, which had been periodically released. It confirmed that the U.S. is "an Arctic Nation with broad and fundamental interests in the Arctic region," seeking to "meet its national security needs, protect the environment, responsibly manage resource[s], account for indigenous communities, support scientific research, and strengthen international cooperation on a wide range of issues."[36]

According to a Congressional report on the Arctic operation, submitted by the Department of Defense in the following year, Thule Air Base would serve as the missile defense system's early warning radar site and play an important role in space missions thanks to the Air Force Satellite Control Network facility there.[37] The report goes on to say that Thule Air Base "supports military, government, and Allied missions in the eastern Arctic."[38] Furthermore, the Department of Defense also could not ignore the effects of climate change. To respond to the melting permafrost and coastal erosion due to uncertain environmental changes, for it became an urgent task to maintain "full mission readiness of the existing early warning radar sites and communications infrastructure in Alaska and Greenland."[39]

The U.S. interests in Greenland, currently, consist of a military objective, scientific research goals, and the Arctic strategy. How these interests were constructed was not necessarily constant throughout the entire period, as they were changing due to the effects of the two surrounding conditions: the security context and the global environment. Although some analysts point out that the U.S. had not shown a proactive attitude as a superpower in the Arctic region (Corgan 2014), at least it may be said that there was hardly any time at which the U.S. interests in any of these aspects receded; rather, the recent trend is the strengthening of its interests there.

---

[34] Brian R. Green (Deputy Assistant Secretary of Defense for Strategic Capabilities, Department of Defense) and Henry A. Obering, Hearing before Strategic Forces Subcommittee, the Committee on Armed Services, House of Representatives, March 27, 2007: 15, 58, 65.

[35] Obering, Hearing before Strategic Forces Subcommittee, the Committee on Armed Services, House of Representatives, March 27, 2007: 65.

[36] The White House, *National Security Strategy*, May 2010: 50.

[37] DoD, *Report to Congress on Arctic Operations and the Northwest Passage*: 18.

[38] Ibid.: 23.

[39] Ibid.: 25.

6 Conclusion

Then, how have the U.S. interests in Greenland changed? From the military point of view, during the Cold War, the importance of the bases in Greenland, including Thule Air Base, was high, as part of the strategy for countering the Soviet Union. It was difficult for the U.S. to find substitutes for them due to their unique geographical positions. After the end of the Cold War, on the one hand, because of the reduction in the military threat from the Soviet Bloc, their strategic value decreased, eventually leading to the withdrawal of troops from Sondrestrom Air Base. On the other hand, at the end of the 1990s, raising threat perception on the missile capability of the so-called "rogue states" accelerated missile defense-related activities of the U.S.

The strategic value of the facilities in the Thule base rose again, but in a different form from that seen during the Cold War. In this process, the function of Thule, Fylingdales, and Alaska became more vital to realizing the U.S. missile defense. However, what is important here is that within the U.S. political process each element was considered not as redundant, but as essential for the overall functioning of the entire "system." Therefore, the more the deployment of the missile defense system progressed globally, the less substitutable each component was. The radar at Thule Air Base was no exception. It seems as if not only Copenhagen but also Washington faces "vulnerability" due to the indispensable values of Thule Air Base in terms of the missile defense project, even though the hypothesis showed in Chapter "Base Politics and the Hold-up Problem" is not directly referred to the U.S. negotiation status.

In considering the military value of Thule Air Base, it is also important to keep watching the behavior of Russia as an environmental factor that can affect not only the military value of the base but also the political opinions within Greenland. Russia has repeatedly protested the deployment of the missile defense system, especially in the European theater. Although the U.S. officially explained that the missile defense shield was intended for repelling attacks from the Middle East, the Rumsfeld Commission Report explicitly expressed the concern on possible Russian missile launches.[40] Russian opposition to the U.S. missile defense shield was not necessarily unreasonable since it at least carried the potential of affecting the Russian military strategy, no matter what the original U.S. intentions were. Taking into account such Russian reactions, Obering even argues that the U.S. should keep informing its posture to Russia.[41] In any case, these gaps in strategic views and interests, detailed

[40] Executive Summary of the Report of the Commission to Assess the Ballistic Missile Threat to the United States.

[41] Obering, Hearing before Strategic Forces Subcommittee, the Committee on Armed Services, House of Representatives, March 27, 2007: 15.

in the next chapter, require certain political consideration both by the U.S. and Greenland.

Military bases in Greenland also play a role as scientific research hubs, which is another important element in the U.S. Arctic interests. This feature, however, affects base politics less directly than do military factors. Certainly, Greenland is highly important for observations of the Arctic environment, and in that sense, it plays a significant role in scientific research infrastructure. The opposition of the U.S. scientific community to the withdrawal from Sondrestrom Air Base would seem to be one piece of evidence for this. At the same time, the function of Thule Air Base as a scientific research hub is expected to be supplemented through international cooperation and by private-sector assets. In other words, although the U.S. scientific research in the Arctic region relies strongly on Greenland, Thule Air Base is not the only venue or means to accomplish this objective.

The U.S. interests in the Arctic region are shaped by the unique geographical position of Greenland, and are currently being more or less realized through Thule Air Base. In addition to that, the U.S. has increasingly focused, especially in the post-Cold War period, on its space operations and the importance of the military and economic activities in the Arctic region, which will further strengthen its interests in the region. The value of Thule Air Base in Greenland, and the Arctic region as a whole, for the U.S. is multi-dimensional (although not necessarily always stable), so it will be, to a certain extent, preserved even if the perceived threats recede and the strategic environment improves. These perceptions and behaviors of the U.S., as "a superior actor", affects the nature of the trilateral relationship, forming a part of the conditions in which the "hold-up" problem emerges.

References

Archer, C. 2003. Greenland, US bases, and missile defence: New two-level negotiations? *Cooperation and Conflict* 38 (2): 143.

Brooke, J. 2000. Greenlanders wary of a new role in U.S. defenses, *The New York Times*, September 18. http://www.nytimes.com/2000/09/18/world/greenlanders-wary-of-a-new-role-in-us-defenses.html?mcubz=1. Accessed 20 Mar 2018.

Peoples, C. 2010. *Justifying ballistic missile defence: Technology, security, and culture*. Cambridge: Cambridge University Press.

Corgan, M. T. 2014. The USA in the Arctic: Superpower or spectator? In *Security and sovereignty in the north Atlantic*, ed. Lassi Heininen, 62–79. Basingstoke: Palgrave Macmillan.

Doel, R. E. 2003. Constituting the postwar earth sciences: The military's influence on the environmental sciences in the USA after 1945. *Social Studies of Science* 33 (5): 635–666.

Fischer, K. 1993. The modernization of US radar installation at Thule, Greenland. *Journal of Peace Research* 30 (1): 7–20.

Kelly, J. 1992. Greenland research projects. *Science* 255 (5052): 1626.

Petersen, N. 2011. Greenland in the U.S. polar strategy. *Journal of Cold War Studies* 13 (2): 90–115.

Reagan, R. 1983. "United States Arctic Policy," National Security Decision Directive (NSDD) 90, April 14. https://fas.org/irp/offdocs/nsdd/nsdd-090.htm. Accessed 20 Mar 2018.

Spring, B. 2000. Clinton's failed missile defense policy: A legacy of missed opportunities, The Heritage Foundation, September 21. http://www.heritage.org/defense/report/clintons-failed-missile-defense-policy-legacy-missedopportunities. Accessed 20 Mar 2018.

Williams, C. J. 2000. U.S. offering Greenlanders a deal they may find hard to refuse, *Los Angeles Times*, July 23. http://articles.latimes.com/2000/jul/23/news/mn-57864. Accessed 20 Mar 2018.

Russia's Military Build-Up in the Arctic: Russia's Threat Perception and its Military Strategy in the Arctic Region

Yu Koizumi

Abstract This chapter explores the background of Russia's recent military build-up in the Arctic region. While those activities have been explained in past studies as Russia's effort to consolidate its control over its natural resources and the Northern Sea Route (NSR) in the Arctic region, this chapter critically examines such understanding and proposes an alternative view. Russia's main motivation is, in our view, based on a more classical military strategy: securing nuclear deterrence.

Keywords Russia · Security · Military strategy · Nuclear deterrence · Arctic · Anti-access and area denial

1 Introduction: The Militarization of the Arctic

Russia's security policy in the Arctic establishes an important context for discussing the various issues considered in this volume. This chapter will explore the strategic implications of the Arctic for today's Russia.

During the Cold War, the Arctic was the frontline of the confrontation between the strategic nuclear capabilities of the United States and the Soviet Union. However, due to the economic turmoil that followed the collapse of the Soviet Union, Russia's strategic nuclear capabilities were greatly weakened. For example, its fleet of nuclear submarines, which had been unable to receive appropriate maintenance, refueling, or supplies, quickly lost its operational capabilities, and by the start of the 2000s, the submarines were in such a condition that they were barely able to perform underwater patrols.[1] The early warning radar and defense networks that extended around the Arctic also stopped functioning and were abandoned, and Russia was only able to maintain a small part of these networks in an active state.

---

[1] Kristensen, H. M. 2013. "Russian SSBN Fleet: Modernizing but Not Sailing Much," *FAS Blog*, May 3, https://fas.org/blogs/security/2013/05/russianssbns/

Y. Koizumi (✉)
The Institute for Future Engineering, Tokyo, Japan

M. Takahashi (ed.), *The Influence of Sub-state Actors on National Security*, Springer Polar Sciences, https://doi.org/10.1007/978-3-030-01677-7_5

However, upon entering the second half of the 2000s, changes in this situation began to occur. In 2007, aerial patrols in the Arctic by strategic bombers, which had been suspended since the end of the Cold War,[2] were restarted, and in the State Armament Program up to 2015 (GPV-2015), Russia decided to develop five of the Type 955 ballistic missile nuclear submarines (SSBN) equipped with the new solid fuel submarine-launched ballistic missiles (SLBM). The number increased to eight in the later GPV. Then in 2011, Russia announced a policy to newly deploy two arctic brigades[3] in Kola Peninsula, which is the location of the base of the Northern Fleet. In addition, airfields and military bases were constructed in the Arctic on the Russian mainland and on islands in the Arctic Ocean. While most of those are radar bases and support facilities for the reconstruction of air defense networks in the Arctic, Russian military has deployed air defense and anti-ship missile units to defend them. In the future, long-range missile systems, such as the K-300P Bastion surface-to-ship missile system and the S-400 air defense system, will be deployed in these bases, and it seems that Russia will also implement the forward deployment of MiG-31 interceptors. Furthermore, every year since 2012, the Russian Navy's Northern Fleet has been conducting large-scale navigation exercises in the Arctic Ocean, as well as actively conducting training exercises in the Arctic.

Considering the organizational aspect, in 2014 the Northern Fleet was given an independent status as the Joint Strategic Command (OSK) "North." Together with the Baltic Fleet, the Northern Fleet had been under the command of the OSK "West," but it was made newly independent and as such took control of the air and ground units within its area of responsibility. The scope of its command is from the Barents Sea coast to the Arctic coast in Siberia,[4] while the eastern side of the Arctic Sea fell under the command of the Eastern Military District (OSK "East").[5]

The question here is what are the intentions of Russia in increasing its military activities in the Arctic? In other words, what is the significance for Russia of the Arctic as a region, and in consideration of this significance, what does it view as threats and how is it attempting to address them? These questions will be examined in the present paper. Depending on the answers to these questions, the responses of

---

[2] According to the Ministry of Defense of Norway, the number of flights by Russian bombers along the Norwegian coast in 2006 was only 14, but this dramatically increased to 88 in 2007 and 75 in 2008. Zysk, K. 2011. "Military Aspects of Russia's Arctic Policy: Hard Power and Natural Resources," in *Arctic Security in an Age of Climate Change,* ed. James Kraska Cambridge: Cambridge University Press: 86. In addition, the patrol flights over the Arctic are conducted by bombers deployed in the Far East, and it has been clarified that in 2014, the flight frequency of these bombers over the Arctic increased by around three times (the exact number is not known). "Воздушный патруль," *Красная звезда,* July 14, 2014.

[3] These units have the ability to conduct strategic military operations in the Arctic using snowmobiles and cold-weather equipment.

[4] *Северный флот.* http://structure.mil.ru/structure/okruga/north/news.html

[5] For example, in "Vostok 2014," which was the large scale war game the OSK "East" conducted in 2014, emergency deployment exercises were conducted on the Chukchi Peninsula by units under the control of the OSK. In addition, there is also a plan to deploy a naval coastal defense fleet on the same peninsula in the future.

the countries in the Arctic (including Denmark, which has sovereignty over Greenland), will be different, and an attempt shall be made to clarify them.

To achieve this objective, first, the strategic significance of the Arctic, and the factors causing this significance to change, shall be investigated. The focus of this paper shall be on answering the questions of how did the Arctic's significance change (or not) for Russia before and after the end of the Cold War, and what were the factors causing these changes. Next, the understanding of this strategic significance of the Arctic for Russia shall be mainly considered through policy documents and the findings of previous studies. In conjunction with this, the actual situation of Russia's strengthening of the military forces it deploys in the Arctic shall also be examined in detail. Finally, the conclusion of this paper shall discuss the prospects of Russia's activities in the Arctic, in terms of how the two aspects of the Arctic, the cooperative and confrontational aspects, are likely to develop in the future.

2 The Arctic's Strategic Significance and the Factors Causing it to Change

(1) The situation since the nineteenth century

Tamnes and Holtsmark were of the opinion that the Arctic's strategic significance was geographical, which they stated was determined particularly by the abundance of resources and the availability of space. In addition to this, both noted that four external factors not directly related to the Arctic itself are responsible for the changes in the Arctic's geopolitical situation. These are "the impulses of humans to discover and explore," "the progress of science," "confrontation and conflict," and "climate change."(Tamnes and Holtsmark (2014)

In other words, the Arctic's strategic significance has come to be determined by the interactions between the absolute value possessed by the Arctic itself (its resources and space) and the external environment that converts these assets into more relative values (of political, economic, and military utility).

What then can be said when applying this sort of framework to security, particularly to classical military security? Before the nineteenth century, the areas in which human activities were possible in the Arctic Circle were extremely limited due to its harsh natural environment. Moreover, prior to the invention of the internal combustion engine, its enormous energy resources were of no value, while exploration and mining technologies were also extremely limited. From a military perspective, while it is not the case that there were no examples of military activities or engagements during this period, they were never strategically epochal to the extent that they determined international relations or the outcomes of wars. To borrow the words of Tamnes and Holtsmark, it can be said that all military events that occurred in and around the Arctic prior to the nineteenth century were nothing more than "minor episodes."

Prior to the nineteenth century, the Arctic's geography had not been clarified and at that stage, rather than for resource utilization and development, expeditions for exploration based on "the human impulses for discovery and exploration" were repeated. This can be considered a stage in which basic findings were accumulated.[6]

(2) The twentieth century: Toward becoming a strategically important region

However, on entering the twentieth century, the importance of the Arctic in terms of security greatly increased. In the First and Second World Wars, the Arctic became a strategic transportation route to supplement the aid supplies to the Russian Empire and the Soviet Union, and battles occurred with Germany that threatened to disrupt this supply route. Despite the existence of restrictions due to the severe natural environment, the Arctic in the northwest part of Russia, centered on Murmansk, was comparatively close to the main warfront in Europe, and its strategic importance was extremely high due to the difficulties created by Germany's blockades of the Baltic and Black Seas. Technological developments had expanded the usable area of the Arctic, and this made it possible to use that land to respond to the requests made to fight the war in Europe.

After the Second World War and upon the start of the Cold War between the United States and the Soviet Union, on the one hand the Arctic lost its value as a strategically important transportation route, but on the other hand, it acquired new value within the context of Cold War strategy. Specifically, the Arctic became significant as the shortest route for the United States' and Soviet Union's bombers and ballistic missiles.[7] The United States' and the Soviet Union's strategic bombers were constantly flying in a state of readiness in the airspace above the Arctic, and a system was in place so that, if war broke out, they could immediately enter the airspace of the other country. This air space was also the flight route for the United States' and the Soviet Union's intercontinental ballistic missiles. This is precisely why the United States and the Soviet Union deployed early warning systems in the Arctic to detect the flights of bombers and ballistic missiles.

After the 1970s, the Soviet Union increased the range of its SLBMs, so it was no longer necessary for it to move them close to the coast of the United States. Although the United States did not move beyond the GIUK gap[8] following the laying of

---

[6] А.Б.Широкорад, *Битва за русскую арктику XVI-XXI вв,* Издательский дом Вече, 2008: 3–92.

[7] The significance of the Arctic in terms of military strategy is summarized in the excellent survey article by Gjert Dyndal of the Ministry of Defense of Norway, Command and Control University. Dyndal, G. 2013. *The Northern Flank and High North Scenarios of the Cold War,* October 24, https://brage.bibsys.no/xmlui/bitstream/handle/11250/285 604/Paper_The%20Northern%20Flank%20and%20High%20North%20Scenarios%20of%20the%20Cold%20War_Dyndal%202,013.pdf.

[8] The GIUK gap refers to maritime space between Greenland, Iceland, and the British Isles. It is a military choke point that Soviet ships had to pass through to travel from the Arctic Ocean to the North Atlantic, and therefore the U.S. Navy laid a severe anti-submarine warning network in this sea area.

anti-submarine networks, it was able to conduct patrols in the Arctic Ocean, which is close to the Soviet mainland. Moreover, the Soviet Union adopted a strategy of strictly protecting the area around the Arctic Ocean with aircrafts, air defense systems, surface vessels, and attack submarines, so the Arctic came to be called a "sanctuary" or a "fortress" by the Western side.[9] The "fortress" was subsequently set as the Sea of Okhotsk, and the SSBNs of the Pacific Fleet were deployed there.

In addition to this, the Soviet Union constructed a nuclear test site on Novaya Zemlya Island in the Arctic, which became a major base for nuclear development. It was also on this island where, in 1961, the Soviet Union detonated "The Tsar Bomb" (yield of 50 megatons), which is the most powerful nuclear bomb ever detonated. Russia has also constructed a Plesetsk satellite launch complex in Arkhangel'sk, which is used for ballistic missile test launches and to launch military satellites.

In this way, the Arctic can be said to have emerged as the front line of the Cold War, owing to the Cold War factors of confrontation and conflict and the progress of science, such as the development of aircraft, missiles, and nuclear weapons. To put this in another way, within the Arctic's absolute value of "resources" and "space," to the latter was added the new significance of being the foundation of nuclear deterrence.

(3) The twenty-first century: The transformation of the Arctic's strategic significance

A new dimension was added to the Arctic in the post-Cold War period. Due to climate change, or the effects of global warming, the ice cap covering the Arctic began to shrink in size. The area of the Arctic's ice cap changes over the course of a year, but in general, this area is biggest in March and smallest in September. In the 1980s, the average ice cap area was 15 million square km in March and seven million square km in September. But according to 2017 data, these areas have shrunk to 13 million square km and four million square km, respectively.[10] Based on historical observations, the ice cap area was the smallest in September 2012 (3.18 million square km), and although this record has not been surpassed in the past 5 years,[11] there has been no change in the fact that its size is trending at a very low level compared to during the Cold War period. It can thus be said that climate change is greatly affecting the strategic significance of the Arctic.

The first example of the Arctic's changing significance is its availability for use as a resource zone. It has long been known that the Arctic has great resource potential, but the political changes following the collapse of the Soviet Union and the end

[9] Dyndal, G. 2017. "50 Years Ago: The Origins of NATO Concerns about the Threat of Russian Strategic Nuclear Submarines," *NATO Review,* March 24, https://www.nato.int/docu/review/2017/Also-in-2017/50-years-ago-nato-concerns-threat-russian-strategic-nuclear-submarines-soviet-bastion-high-north/EN/index.html

[10] The National Institute of Polar Research, *ADS (Arctic Data archive system).* https://ads.nipr.ac.jp/vishop.ver1/ja/vishop-extent.html

[11] The National Institute of Polar Research. 2017 "The Area of Sea Ice in September in the Arctic Ocean: The Record Smallest Area Was Not Observed for the First Time in 5 Years," September 15, http://www.nipr.ac.jp/info/notice/20170915.html

of the Cold War (the easing of confrontation and conflict), and the progress of science in terms of the development of exploration and mining technologies, has further expanded the availability of resources for exploitation in the Arctic. According to a 2008 report by the United States Geological Survey (USGS),[12] in terms of undiscovered resources, the Arctic has about 90 billion barrels of oil, 1760 trillion cubic feet of natural gas, and 44 billion barrels of liquid natural gas, which can be said to be equivalent to 13% of the undiscovered oil in the world and 30% of the undiscovered natural gas (78% if limited to liquefied natural gas).

Second, for the same reason as the increased availability of resources for use, the availability of the Arctic Ocean as a sea route has expanded. In terms of the sea route from Europe to the Far East (for example, to Japan), while the route through the Indian Ocean via the Suez Canal (approximately 15,000 nautical miles) is currently being used, by using the Northern Sea Route (NSR) around the coast of Russia, it is possible to decrease the length of this voyage to around 7000 nautical miles, or less than around half of the distance of the currently used route. Furthermore, in the case of connecting Asia to the east coast of North America, it is estimated that if the northwest route around Canada is taken, which is 7000 nautical miles, the distance travelled can be reduced by around 30% compared to the current route of crossing the Pacific (around 10,000 nautical miles). Previously, it was generally impossible to use Arctic sea areas because their surfaces were covered with ice, with passage only being possible for icebreaker vessels or a fleet escorted by an icebreaker. However, due to the reduction in the ice cap area, there are times during the summer in which the NSR is not covered by ice. In addition, the multi-year ice that exists throughout the year has been compressed for many years, and as a result, it is extremely dense and difficult to break. However, newly formed first-year ice has low density, and therefore the sea areas where the ice disappears in summer are easier to navigate– even in winter–than in the past.(Huebert 2017)

The Soviet Union opened the NSR as an international route in 1991; however, the route only began to be used in 2009. This is because the effects of global warming only became conspicuous upon entering the twenty-first century; moreover, the possibility of using the NSR to reduce navigation distances and thereby save on fuel costs only attracted attention following the spike in the prices of international energy resources.[13] The Russian government also revised its law on commercial navigation in 2012 and established a law for the commercial navigation of the Arctic Ocean. In addition, it began to establish a search and rescue system to ensure the safe passage of vessels.[14] The utilization of the Arctic Ocean as a sea route steadily increased

[12] *Circum-Arctic Resource Appraisal: Estimates of Undiscovered Oil and Gas North of the Arctic Circle*, U.S. Geological Survey, 2008, https://pubs.usgs.gov/fs/2008/3049/fs2008-3049.pdf

[13] Kramer, A. E. 2011. "Warming Revives Dream of Sea Route in Russian Arctic," *New York Times*, October 17, http://www.nytimes.com/2011/10/18/business/global/warming-revives-old-dream-of-sea-route-in-russian-arctic.html

[14] Koizumi, Y. 2014. "Maritime Legislation in Russia: Focusing on Security Policy in the Arctic Ocean" *Foreign Legislation*, no. 259: 89–91.

until the middle of the 2010s, and in 2013, a record 71 vessels passed through the NSR with a total cargo volume of 1.35 million tons.[15]

On the other hand, the Arctic did not lose its former significance in terms of its role in Russia's nuclear strategy. After the collapse of the Soviet Union, Russia's military forces fell into a state of major confusion, as was touched upon at the start of this paper, and therefore Russia considered its nuclear weapons to be even more important as the last resort for ensuring its security. In other words, it can be said that the receding of the confrontation and conflict factor did not lower the Arctic's value as the foundation of Russia's nuclear deterrence. Incidentally, according to Russian military commentator Alexander Golts, the Russian Navy adopted the strategy of turning the Arctic Ocean into a SSBN "fortress" in 1992. (Golts 2011) If this claim is correct, although the Soviet Union's fortress strategy was observed by the West during the Cold War, it was not a very clear concept at that time and was instead only adopted in the post-Cold War period.

However, climate change, which has made economic activities in the Arctic possible, is having a major impact on this approach to the Arctic as a fortress supporting nuclear deterrence. Due to the reduction in the ice cap area, the SSBNs, which were previously protected by the ice, have become more vulnerable; moreover, there has emerged the possibility of the United States deploying in the Arctic Ocean its Aegis vessels that are capable of intercepting and destroying ballistic missiles (This is described in detail in the next section).

Above, we saw that there have been changes due to the factors of climate change and scientific progress, but there are also new changes due to the confrontation and conflict factor: the level of military tension in the Arctic is rising due to the worsening and straining of relations between Russia and the West. President Putin in his first term sought to improve relations with the West, and in the Arctic a good cooperative relationship was established between Russia and the West in areas such as the dismantling of retired nuclear submarines and environmental protection.[16] However, relations between the United States and Russia have gradually worsened since then over issues such as the Iraq War, the deployment of the missile defense (MD) system in Eastern Europe, and the democratization of former Soviet countries; the effects of this strained relationship have also extended to the Arctic. As previously mentioned, Russia restarted bomber patrols in the Arctic, a decision that it made in this context of strained relations with the West. Since then, Russia has been working to strengthen its military forces, including its nuclear capabilities, in the Arctic, and this development was further accelerated by the Ukrainian crisis of 2014.[17]

---

[15] Centre for High North Logistics Information Office, *NSR Transit 2013*, http://www.arctic-lio.com/docs/nsr/transits/Transits_2013_final.pdf

[16] Huebert positioned the period from the end of the Soviet Union in 1989 to 2006 as the peak period for cooperation on the Arctic. Huebert. "Security Challenge": 369.

[17] Although the war in Georgia in 2008 and the Ukrainian crisis from 2014 onwards were themselves low-intensity conflicts, Russia indicated the possibility that they would be escalated by interventions from the West. Therefore, it is believed that in Russia's military strategy, the deter-

3 The Importance of Arctic for Russia

(1) The Arctic's economic value and the rise in its threat perceptions

Based on the Arctic's strategic significance outlined in Sect. 2, in this section we will consider how Russia positions the Arctic within its security policy.

Currently, Russia's Arctic policy is most comprehensively defined by the "Basic National Policy on the Arctic for the Period of 2020 and Beyond" (hereafter, "the Basic Arctic Policy")[18] that it formulated in 2008. The foundation of the Basic Arctic Policy is reflected deeply in the environmental changes that took place in the Arctic in the post-Cold-War period described in Sect. 2(3).

The first point is that Russia has positioned the Arctic as the "Russian Federation's strategic resource base." As already mentioned, the Arctic potentially has enormous reserves of natural resources, and the development of these resources is of critical significance for Russia if it is to secure its position in the future as a world power. This is because the main resource zones that supported the Russian economy in the past have already exceeded their production peaks,[19] so it must develop new energy resource production areas.[20]

Second, Russia positions the NSR as the "Federation's national, unified transportation route in the Arctic," and it intends to maintain and use it on a national scale. As has already been briefly touched upon, for Russia, the significance of the NSR as a route for vessels is not limited to it simply being a transportation route. This is because if the NSR became a global trade route, like the Indian and Atlantic Ocean routes, there is a possibility that "Russia would be raised from the periphery to the center of the global economic system."[21]

Based on the above-described awareness, in the Basic Arctic Policy, Russia has set its international priority goals for the future as expanding the use of the Arctic Circle as the foundation of its strategic resources, enhancing the defense and border security system of the Arctic Circle, protecting the ecosystem, establishing an

rence against such an escalation through nuclear capabilities was integrated into its strategy for small-scale conflicts. Please refer to the following for more details: Koizumi, Y. "Russia's Intervention Strategy as Seen from the Ukrainian Crisis: What is its Hybrid Strategy?" *International Issues*, No. 658: 38–49.

[18] *Основы государственной политики Российской Федерации в Арктике на период до 2020 года и дальнейшую перспективу*, http://scrf.gov.ru/security/economic/document98/

[19] At a security conference held in December 2010 on the theme of energy security, it was reported that 50% of the confirmed reserves of Russia's oil resources had already been excavated, and that the remaining confirmed reserves of natural gas were 165 trillion cubic meters. "Энергетика социальной безопасности," *Российская газета*, December 14, 2010.

[20] Please refer to the following for the current state of Russia's resource development in the Arctic Circle. Harada, D. 2017. "Under the European and US Sanctions: The Current State of Russia's Development of Oil and Gas in the Arctic Circle," *Oil and Natural Gas Review*, 51, no. 5: 43–78.

[21] Markushina, N. and M. Lagutina. 2017. "The Arctic Region and the 'New North': The Russian Approach," in *Russia and the World: Understanding International Relations*, ed. Natalia Tsvetkova. Lexington Books: 332.

information communication network, advancing scientific research, and promoting international cooperation.

In 2013, Russia approved the Russian Federation's Strategy for Development of and Security in the Arctic Circle in the Period up to 2020 (hereafter, "the Arctic Strategy")[22] in a form that embodied the Basic Arctic Policy. The Arctic Strategy sets the targets of securing the resources contained in the Arctic Ocean's continental shelf, the year-round use of the Arctic Ocean sea route, and connecting this route to rivers, railways, and the air network to use it as a multi-faceted route for exporting resources, and it can be said to basically follow the same line as described in the Basic Arctic Policy. It also provides for the establishment of a fleet of icebreaker vessels toward the year-round use of the NSR.[23]

As described above, Russia has found national significance in the Arctic's economic usefulness. However, at the same time, this is also causing the Russian government's threat perceptions of the Arctic to rise. For example, the Russian Federation National Security Strategy up to 2020,[24] which is a comprehensive security policy document published in 2009, outlines the prospect of the potential competition between states over the Arctic's resources as follows: "The focus of international politics in the long-term future will be on the control of natural resources in the Middle East and on the continental shelf in the Barents Sea and other Arctic regions, as well as in the Caspian Sea and Central Asia." Furthermore, based on this prospect, the government sets as priority issues the strengthening of border protection around Russia, including in the Arctic, and establishing and maintaining economic, border, and military infrastructures. In the successor policy document to this document, the Russian Federation's National Security Strategy,[25] the current version of which was published in 2015, it indicates the following awareness: "Confrontations are intensifying over imbalances in global development, the widening of the disparity in the levels of wealth between countries, conflicts over resources, access to resource markets, and control over the main transportation routes. In the competition between nations, in whichever the case, the parts relating to social development values and models are dominant. In this process, who has leadership in the resource development in the world's oceans and in the Arctic is of great importance."

In the previously described Arctic Strategy, the security aspect, which is described below, greatly increased in importance compared to the Basic Arctic Policy. It notes the necessity of military forces in the Arctic to protect economic activities, such as

---

[22] *О Стратегии развития Арктической зоны Российской Федерации и обеспечения национальной безопасности на период до 2020 года*, http://government.ru/info/183 60/.

[23] The Soviet Union built six nuclear icebreakers for the navy, and four of these are now operated by the state-run Atomflot. In 2017, the first new-type nuclear icebreaker, the successor of the former vessels, was also launched.

[24] *Стратегия национальной безопасности Российской Федерации до 2020 года*, https://rg.ru/2009/05/19/strategia-dok.html

[25] *Стратегия национальной безопасности Российской Федерации*, http://www.scrf.gov.ru/security/docs/document133/

enabling activities in exclusive economic zones and on the continental shelf to prevent military pressure and invasions in the Arctic. It also includes the deployment of troops to defend the Arctic and the establishment of the related infrastructure.

Not limited to policy documents, this sort of awareness can frequently be seen in the comments of senior Russian government officials. For example, in 2013, Deputy Prime Minister Rogozin (known for comments on defense as the former ambassador to NATO) commented that resource development on the Arctic's continental shelf will inevitability lead to conflict between countries and that there may be sabotage against Russia's oil and gas mining activities by foreign countries.[26] In addition, in 2015, Director of the Federal Security Bureau (FSB) Bortnikov, who is also the Secretary of the National Counter Terrorism Committee, identified the need to strengthen counter-terrorism measures in response to the increasing resource development of and passage by vessels through the Arctic.[27]

However, as has already been noted, absolute value, such as resources and sea passages, should not directly lead to conflict. As seen in the confrontation between Russia, Canada, and the various countries of Europe over the extension of the Arctic continental shelf, although it is true that conflicts exist over securing the Arctic's resources, resource development is not necessarily a zero-sum game. If viewed from Russia's perspective, Europe and the United States have the essential elements for the development of the Arctic's resources, and the necessary funds for exploration and mining technologies. Thus, diplomatic solutions are possible that will serve the interests of both sides, as was seen in the maritime border demarcation of the Barents Sea (2010). At the very least, conflict over resources cannot be said to legitimize the rapid militarization that Russia is conducting.

Alternatively, as mentioned by Korzun on the conflicts over ocean resources, conflicts occur as the sum of "conflict situations + causes."[28] Therefore, it would seem necessary to pay attention to situations in which conflict conditions can change into actual conflicts.

Thus, a question emerges: what kinds of external conditions can convert conflict conditions into conflicts?

What we can first point to is the confrontation and conflict factor in U.S.-Russia relations. The Putin administration, which came to power in 2000, initially tried to improve relations with the United States through cooperation in areas such as the war in Afghanistan, nuclear arms control, and MD, as well as by accepting Baltic countries' membership in NATO. However, relations between the United States and Russia became strained because of issues such as Russia's opposition to the war against Iraq, former Soviet Union republics' approaches to EU and NATO. Russia decided to resume aerial bomber patrols in 2007 in this international political

---

[26] "Рогозин: против объектов РФ в Арктике могут быть совершены диверсии," *РИА Новости,* 2013. 6.28.

[27] "Арктику обезопасят от террористов: Александр Бортников встал на защиту Северного морского пути," *Независимая газета,* August 12, 2015.

[28] В.А.Корзун, *Конфликтное использование морских и прибрежны зон россии в XXI веке,* Экономика, 2004: 6.

climate, and the confrontation between the two countries reached a peak in 2008 upon the outbreak of war in Georgia.

Immediately after that war, the following argument was developed in a paper entitled "Strategic Deterrence: A New Concept for Russia's Military Security,"[29] which was published by retired General Makhmut Gareev, a well-known thinker of strategy. In it, he noted that global interdependence is progressing for problems on the scale of humanity, of economic, environmental, energy, societal, political, and psychological problems, and no country could solve such problems alone. However, contrary to this development, the United States was trying to gain "unipolar control" by positioning itself as the "world government" - however, this approach consisting of a mass consumption economy, democracy with low diversity, and the unilateral exercise of military force was reaching its limits. In this situation, Gareev emphasized that the United States, rather than waging wars through the direct exercise of military power, was waging "wars that are not seen as wars" though the utilization of politics, economics, thought, psychology, information, intelligence, counter-intelligence, and counter-terrorism activities.

This sort of anti-American worldview, like the conspiracy theories during the Cold War, has had a major influence on Russia's military thinking. In 2013, Chief of the General Staff Gerasimov, who is the head of the "uniforms" of the Russian army, in a paper entitled, "The Value of Science in Forecasts,"[30] developed the argument that, keeping in mind the series of political changes that occurred in the Middle East in the 2010s (the so-called Arab Spring), the "wars not seen as wars" that are being conducted by the United States are the typical form of the armed conflict in the twenty-first century. In addition, Enduring Bonds 2013, joint exercises by the Collective Security Treaty Organization (CSTO) conducted in the same year, were implemented for a scenario in which a fictional state, "the Ural Republic," is attacked by anti-government forces with the backing of a "powerful military and political alliance" that aimed to "control its energy resources."[31] Since then, the Russian army has frequently used the scenario of "wars that are not seen as wars" by the West in its large-scale exercises.

(2) A classical perception of military threat: Nuclear deterrence and the US Missile Defense (MD)

However, this sort of anti-American sentiment increased from the end of the 2000s to the 2010s, so it cannot explain all of Russia's threat perceptions about the Arctic before then (or in the long-term future). For this, it would seem necessary to point to a more purely military aspect.

As stated by Zysk, the prevailing thought is that Russia's military strategy tends to be strongly based on the opponents' capabilities rather than on its intentions, and

---

[29] М.А. Гареев, "Стратегическое сдерживание- новая концепция военной безопасности России," *ИА REGNUM*, 2008.10.8.

[30] Валерий Герасимов, "Ценность науки в предвидении," *Военно-промышленный курьер,* No. 8 (47 6), February 27, 2013.

[31] "Миротворцы ОДКБ защитили "Республику Уралия," Российская газета, October 8, 2013.

as the futures of the external factors relating to absolute value are uncertain, it is necessary to prepare for this uncertainty.[32] Considering that all four countries, other than Russia, with territory in the Arctic Ocean are NATO member states, dispelling Russia's threat awareness is unlikely to be easy.(Sergunin and Konyshev 2015) If seen from Russia's perspective, it is the NATO countries that are advancing militarization and creating the threat in the Arctic.[33] In the U.S. Navy Arctic Roadmap from 2014 to 2030,[34] the United States' Department of Defense set out its measures to improve its rapid response system and patrol activities in the Arctic Ocean, and Russia has indicated that it feels a strong sense of threat regarding these measures.[35] Furthermore, President Putin has said that the Arctic contains important resources and is a nuclear deterrent, so Russia must be wary of the capabilities of the U.S. Navy in the Arctic.[36]

In addition, the previously mentioned reduction in the area of the ice cap due to climate change is spurring Russia's threat perceptions. The fact that the Arctic Ocean was a "sanctuary" for the Soviet Union's and Russia's SSBNs is not only because Russia deployed defensive forces around it, but also to a large extent because the sea surface was covered with thick ice. Because of this, enemy anti-submarine warships and aircraft could not operate in the Arctic Ocean,[37] and the threat to SSBNs was limited to that posed by the enemy's attack submarines that could operate over long distances. However, the reduction in the area of the ice cap due to global warming has destroyed this precondition. Particularly during the summer, it has become fully possible for conventional anti-submarine warships and aircraft to operate in the Arctic Circle, so it is expected that the "fortress" of the sea areas where Russia's SSBNs can patrol will have to be shifted to locations nearer the North Pole, to where it is harder to provide support.

Moreover, the threat from climate change is not limited to SSBNs. This is because, as the Arctic Ocean can be navigated by surface vessels, the possibility has arisen that Russia's inland areas will be hit by simultaneous precision cruise missile strikes launched from the Arctic Ocean, and that the retaliatory attack in response to these strikes will be intercepted by the United States' MD system installed on the Aegis vessels.[38] What would seem to be particularly important here is the

[32] Zysk, "Military Aspects of Russia's Arctic Policy": 99–100.

[33] This type of threat awareness is seen in practically all the Russian works on Arctic security. For example, please refer to the following. А.И. Подберезкин, *Военные угрозы России,* Издательство «МГИМО-Университет,» 2014: 130–131.

[34] *U.S. Navy Arctic Roadmap 2014–2030*, http://greenfleet.dodlive.mil/files/2014/02/USN-Arctic-Roadmap-2014.pdf

[35] Sergunin and Konyshev, *Russia in the Arctic*: 145.

[36] "Putin Says Russia Needs to Beef up Arctic Presence," *AP,* May 12, 2013.

[37] This is because anti-submarine vessels cannot navigate the ice-covered sea area and anti-submarine aircraft cannot drop sound buoys under the ice.

[38] А. А. Храмчихин, "Станет ли Арктика театром военных действий по последнему переделу мира?" *Арктика и Север,* no.10, 2013, 2.; Юрий Банько, "Арктика – зона конфронтации: Россия вступает в битву за свои интересы на Севере," *Независимое военное обозрение,* April 18, 2014.

deployability of the latter, the MD system. The possibility of a pre-emptive strike by cruise missiles (a strategic attack to disarm nuclear capabilities) is a problem that has been discussed since the Cold War, but since the deployment of the cruise missile launch platform takes some time, a complete surprise attack is impossible. Moreover, there is the argument that due to their slow flight speed, there remains sufficient room for a retaliatory attack if they are unable to simultaneously strike all of Russia's nuclear capabilities (therefore, a pre-emptive attack by cruise missiles can be deterred).[39] However, there has arisen the new possibility that such disarmament strikes and the MD system will be deployed at the same time. Moreover, considering that the flight course of ballistic missiles is over the Arctic, the degree of threat for Russia's nuclear deterrence is likely to be higher than that from an Eastern European MD. This assumption leads to the revision of the strategic value of the Greenland. In the computer simulation conducted by Russia's General Staff, ballistic missiles launched from the central Russia shall be, at first, detected by the U.S. early warning satellite, then tracked by Sea-Based X-band Radar (SBX-R) deployed in the northern Pacific and early warning radar installed in Tule.[40] This means that the U.S. presence in Greenland is understood as a component of the global missile defense network.

Furthermore, as shown by the estimates of Pyryev and Dvorkin, if ICBMs are launched from the main missile units in Western Russia (Vypolzovo, Tatishchevo, and Dombarovskiy) to the east coast of the United States, depending on the speed and the flight angle of the missile, it is possible that Aegis vessels deployed near the GIUK gap will be able to intercept them. In March 2015, Mikhail Vanin, the Russian ambassador to Denmark posted an article to the Danish newspaper "Juras Bosten," in which he argued that "Danish warships shall become the target of Russian nuclear missiles," and "Denmark will lose money and security," if the country joins the U.S. led-European missile defense. This argument could be interpreted as an expression of Russia's concern over Denmark's and Greenland's strategic devotion to the Arctic deployment of missile defense systems.

However, as Pyryev and Dvorkin recognize, Russia's strategic nuclear capabilities are significant, so the redundancy of pre-emptive strikes is high. For example, even if 80% of the ICBMs and 50% of SSBNs were destroyed by disarmament strikes by the United States, and 100 interceptors were launched against the Russia's retaliatory attack, around 450 of its strategic nuclear warheads could still reach the U.S. mainland. Therefore, they have concluded that the deployment of Aegis vessels in the Arctic Ocean would not considerably damage Russia's nuclear deterrence against the United States.[41]

---

[39] Владимир Дворкин, "Сдерживание и стратегическая безопасность," Алексей Арбатов, Владимир Дворкин ред., *Ядерная перезагрузка*, Московский центр Карнеги, 2011: 23–45.

[40] В.В. Герасимов, *Оценка глобального потенциала ПРО часть 3*, 2012.5.5. lessthanhttp://mil.ru/files/morf/2-2_m2%20centr.wmvgreaterthan

[41] Владимир Пырьев, Владимир Дворкин, "Программа США/НАТО и стратегическая стабильность," Алексей Арбатов, Владимир Дворкин ред., *Противоракетная оборона: Противостояние или сотрудничество?* Московский центр Карнеги, 2012: 173–191.

Nevertheless, as the threat level against its SSBNs rose and it became possible that the missile defense system would be deployed, Russia needed some measures to counter this increased threat. It is believed that Russia is investigating deploying in the Arctic ground-to-ship missiles and air defense assets with the aim of building A2/AD capabilities (anti-access / area denial) to prevent and impede the activities of U.S. anti-submarine units and missile defense units in the Arctic Ocean.

Russia is also building A2/AD capabilities in locations such as the Baltic Sea, the Black Sea, and the East Mediterranean Sea, but these are somewhat different in character to the A2/AD network in the Arctic. All of these are regions where Russia conducts military interventions or may do so, and A2/AD exists to defend Russia's intervention forces (mainly from a counter-intervention by the West). On the other hand, the A2/AD network in the Arctic Ocean is intended to defend the strategic nuclear deterrent, and in this sense, its commonality with the construction of an A2/AD network in the Sea of Okhotsk is high. In this sense, within Russia's strategic thinking, the Arctic Ocean and the Sea of Okhotsk are considered to be a single area confronting the United States' threat to its nuclear deterrent.[42]

Moreover, Russia is also advancing the modernization of the SSBNs that play the fortress role. For the previously mentioned Type 955 SSBN, it has already completed the construction of three vessels, with one being deployed to the Northern Fleet and two to the Pacific Fleet. It has also begun the work of constructing all the remaining five vessels and it is believed that finally, four of the Type 955 submarines will each be deployed to the Northern Fleet and Pacific Fleet. In addition to this, Russia has completed the modernization of six of the Tyep 667BDRM SSBN constructed during the Soviet era (all in the Northern Fleet) by installing in them the new Sineva SLBM, and it is thought that they will remain in active service up until the 2030s.

(3) Russia's status as a world power

It has frequently been noted that the intention of Russia's Arctic policy is to try to increase its political influence and, especially, to secure its position as a world power. As pointed out by Markushina and Latutina, Russia is trying to acquire the status of a world power in the new international order, supported by the rising international importance of the Arctic issue, and also aiming to secure its leadership among the Arctic countries. Conversely, if the Arctic countries seek to cooperate not with Russia, but with Western countries and China, this would be a major setback from Russia's perspective.[43]

In terms of security, which is the theme of this chapter, since the Ukrainian crisis, the tendency of Sweden and Finland to draw closer to NATO has been increasing

---

However, in a separate single-author article included in this book, Dvorkin states that if the missile launched is a liquid fuel-type SLBM with a slow rate of ascent, it may be intercepted by missiles developed in the future. Владимир Дворкин, "Перспектив противоракетного сотрудничества США/НАТО и России," *ibid,*: 197–198.

[42] Sergunin and Konyshev, *Russia in the Arctic*: 145–146.

[43] Markushina and Latutina, "The Arctic Region": 327–328.

like never before. Currently, even though this still remains in the realm of possibility, if both of these countries become members of NATO, the balance of power in Northern Europe that has been maintained up to the present between Russia and the United States (the so called "Nordic Balance") will be greatly shaken, which would likely further strengthen Russia's perception that its security is threatened from all sides. Some believe the increase in Russian military activities in the Arctic will keep such a situation in check,[44] but as stated by Huebert, it is also highly possible that it will have the opposite effect.[45]

4 Conclusion: The Result of Strategic Miscommunication

To summarize the discussion above, Russia's military build-up in the Arctic could be viewed as a result of strategic miscommunication between the U.S. and that country.

Post-Cold War Russia's strong frustration over the U.S. unilateral military interventions, missile defense system deployment that ignored Russia's opposition, democratization of the Post-Soviet countries, etc. led to the Ukrainian crisis in 2014. This series of events did not undermine Russia's military security directly, and did not raise the probability of a Cold War-like large-scale state-to-state war, as Russia's security policy documents, including the Military Doctrine, have frankly admitted. However, the deterioration of the U.S.-Russian relationship has spurred Russia's tendency to "prepare for the enemy's capabilities, rather than intentions." Under such circumstances, Russian military policy has been increasingly relying on the nuclear forces to maintain military deterrence against the U.S. Repeated U.S. explanations aimed at appeasing Russia have not succeeded.

Moreover, such Russia's behavior has produced a vicious cycle, provoking NATO members' concerns. While the countries of the West understand that the probability of Russia's direct military intervention in the Arctic is quite low because Russia's main interest is preserving its sphere of influence in the Post-Soviet space, they cannot help being alarmed in the face of Russia's military build-up. Especially, the modernization of Russia's nuclear capabilities has raised concerns in NATO countries and made them revisit the strategic value of Greenland.

Herein lies the paradox: the strengthening of military capability itself invites counter-militarization.

---

[44] "Russia Warns Sweden and Finland Against NATO Membership," *Defense News*, June 12, 2014.

[45] Huebert, "Security Challenge": 365.

References

Golts, A. 2011. The Arctic: A clash of interests or clash of ambitions. In *Russia in the Arctic*, ed. Blank, S. J., 53–54. Pennsylvania: Strategic Studies Institute.

Huebert, R. 2017. Security challenge in the Arctic. In *Global challenges in the Arctic region*, ed. Conde, E. and S. I. Sanchez, 366. New York: Routledge.

Sergunin, A. and V. Konyshev. 2015. *Russia in the Arctic: Hard or soft power?* Vol. 145. Stuttgart: ibidem Press.

Tamnes, R. and S. G. Holtsmark. 2014. The geopolitics of the Arctic in historical perspective. In *Geopolitics and security in the Arctic*, ed. Tamnes, R. and K. Offerdal, 13. New York: Routledge.

Okinawa's Search for Autonomy and Tokyo's Commitment to the Japan-U.S. Alliance

Shino Hateruma

Abstract Greenland and Okinawa, the southernmost island prefecture of Japan, are both, as sub-state entities, involved in triangular interactions with their central governments and the U.S. over issues pertaining to American military bases. Examining the case of Okinawa provides us with a useful reference point for considering the case of Greenland because of the similar characteristics Okinawa exhibits in terms of U.S. presence and center-periphery relationship. Considering that Okinawan autonomy eroded through its annexation by Japan, involvement in the Second World War, and the 27-year occupation by the U.S. military, how did this sub-state actor seek autonomy, come into conflict with the national security of Japan, and end up being restrained by the center? After the end of the Cold War, Okinawa challenged the continuing U.S. presence, thus endangering the foundations of the security relations between Japan and the U.S. This chapter argues that the people of Okinawa raised their voices and protested U.S. military presence in order to be able to handle their property and future development on their own, and that, while they were able to effect partial changes in the locations and activities of U.S. bases, they were not able to significantly alter U.S. force posture in the islands. The main reason for this is that the central government in Tokyo limited Okinawa's influence on national security policy by exerting its legal and administrative power.

Keywords Okinawa · U.S. bases · Japan-U.S. alliance · Sub-state entity · Anti-base protest · Autonomy

S. Hateruma (✉)
Graduate School of Asia-Pacific Studies, Waseda University, Tokyo, Japan
e-mail: sthateruma@toki.waseda.jp

M. Takahashi (ed.), *The Influence of Sub-state Actors on National Security*, Springer Polar Sciences, https://doi.org/10.1007/978-3-030-01677-7_6

1 Introduction

The thinking regarding the manifestation of sub-state actors' aspirations for self-rule (intention) and actions in the domain of national security that is analyzed in this book through the case of Greenland can potentially be applied beyond the Arctic region. As Alexander Cooley points out, regarding the triangular interactions involving U.S. bases abroad, Greenland and Okinawa behave as actors that negotiate with central governments and the U.S., rather than leaving their national authorities to settle issues pertaining to U.S. military bases through bilateral arrangements with Washington.[1] This characteristic derives from their similar situations and backgrounds regarding the U.S. bases. Both Okinawa and Greenland have significantly unbalanced and concentrated U.S. presence compared to their respective mainlands. Although, unlike mainland Denmark, mainland Japan does provide several bases for U.S. Forces in Japan (USFJ). Despite the difference in status—Greenland as an autonomous Danish territory and Okinawa as a Japanese prefecture—both have historically been tied to the central government as part of a hierarchical relationship. The case of Okinawa, which has similar characteristics as Greenland in terms of U.S. presence and center-periphery relationship, deepens understanding of Greenland's exertion of autonomy and its impact on national security policy.

Okinawa is located between the Japanese mainland and the Philippines. Major cities in East Asia—Tokyo, Seoul, Shanghai, Hong Kong, and Manila—are within about a 1500 kilometer radius from Naha, the prefectural capital of Okinawa. The prefecture is composed of 160 islands and islets extending 1000 kilometer east-west and 400 kilometer north-south. 70.6% of USFJ facilities and areas, which are used exclusively by U.S. military forces, are located in Okinawa alone. U.S. bases occupy about 15% of the Okinawan main island, on which more than 1.3 million people live. The manpower is also concentrated in Okinawa, which accommodates 25,843 of the 36,712 U.S. personnel stationed in Japan.[2] All services are based in the prefecture: the Army's communication site; the Navy's berthing and storage areas; one of the Air Force's largest air bases in the Far East with ammunition and fuel storage; and the Marine Corps' two airfields, logistic facilities, and training areas with helipads. Among them, the Marine Corps' presence is the largest in terms of the size of the bases and the number of troops. The Japanese government, which has allowed large-scale U.S. deployment in the area, notes that Okinawa's strategic location enables U.S. forces to respond to various contingencies, support the functioning of the Japan-U.S. alliance, and contribute to the security of Asia and the Pacific.[3]

---

[1] Cooley, A. *Base Politics: Democratic Change and the U.S. Military Overseas*. Ithaca, NY: Cornell University Press, 2008: 137–174.

[2] The number of personnel is as of June 2011. Okinawa-ken. Okinawa no beigun oyobi jieitai kichi [Bases of U.S. Military Forces and the Japanese Self-Defense Forces in Okinawa]. Naha: Okinawa-ken 2017 (In Japanese); "U.S. Military Base Issues in Okinawa," Okinawa Prefectural Government Washington D.C. Office. Retrieved from http://dc-office.org/basedata#p3

[3] Japan Ministry of Defense. 2017. *Defense of Japan*: 288

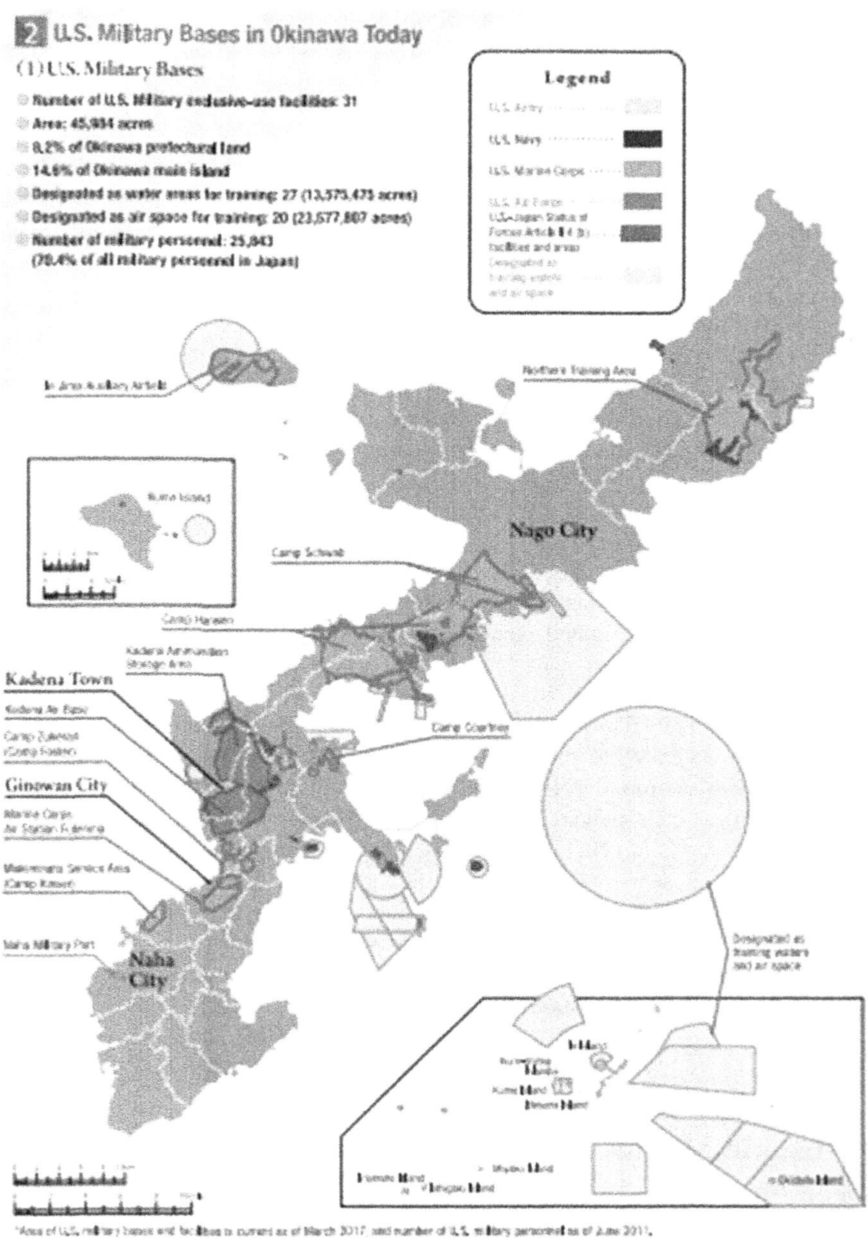

Source: Okinawa Prefectural Government Washington Office. http://dc-office.org/wp-content/uploads/2018/03/E06.pdf

In accordance with the main theme of this book, autonomy and its impacts on national security, this chapter examines how the sub-state actor Okinawa has sought to develop its autonomy, clashed with Japanese national security interests, and been restrained by the center. The clash of interests was triggered by an incident in which three U.S. servicemen assaulted a schoolgirl in Okinawa in the fall of 1995. This chapter argues that Okinawa, in search of autonomy, protested U.S. military presence, and that, while it successfully brought about changes in the locations and activities of some U.S. bases, it could not significantly alter the posture of U.S. forces on the island due to the U.S. deployment strategy and the overarching Japan-U.S. alliance being readjusted to accommodate the post-Cold War security environment. Moreover, Okinawa's protest had the opposite effect. Its right of veto over U.S. military bases was neutralized through legal and administrative procedures to ensure the stable provision of land for the military bases. In short, unlike Greenland, Okinawa lost its leverage in directly influencing national security policy, but continued its commitment to the alliance by conceding basing rights to the U.S.

To understand the background of Okinawa's motivation for autonomy, this chapter first reviews the trajectory of the erosion of its autonomy. Although once independent, Okinawa later endured the rule of the Empire of Japan and then the U.S. During that period of control by outsiders, the people of Okinawa struggled for autonomy. This quest continued even after Okinawa was returned to Japan, because of the concentrated presence of U.S. forces. The following section deals with domestic and international affairs regarding U.S. bases in Okinawa in the 1990s, when the prefecture challenged the constant stationing of U.S. troops that were seen as having the potential to endanger security relations between Japan and the U.S. The third section describes the international circumstances surrounding Okinawa and the U.S.'s view of its military deployment in East Asia. The discussion is based mainly on U.S. official documents, such as strategic reports, with the goal of clarifying the subsequent persistence of the Japan-U.S. alliance in the mid-1990s. The fourth section examines the clash between Okinawan autonomy and alliance interests, as well as how Japan thwarted Okinawa's attempt to increase autonomy. The interaction between Tokyo and Okinawa is brought to light by investigating Japanese Diet minutes, memoirs of relevant decision makers, and newspaper articles.

2 Historical Background of Okinawa's Quest for Autonomy

2.1 From an Independent State to a Battlefield

Pre-modern Okinawa was an independent entity called the Ryukyu Kingdom from 1429 to 1879. Under the Chinese tributary system, Ryukyu flourished as a "cornerstone of East Asia" through trade with China, Japan, and Korea to the north and

Annam (Vietnam), Siam (Thailand), and Java to the south.[4] In 1609, the Satsuma Domain, a then-feudal fief located on the Japanese island of Kyushu, invaded Ryukyu. Satsuma gradually imposed nationwide obligations, such as tax and trade control, on Ryukyu as part of a fait accompli strategy of keeping it under control.[5] Meanwhile, as a "dual attachment to both Japan and China," the Ryukyu Kingdom enjoyed political, economic, and intellectual autonomy.[6] However, Japan officially annexed Ryukyu in 1879, an act known as the Disposition of Ryukyu, and embarked on a gradual dissolution of Ryukyuan customs such as those related to the land and the language.[7]

After winning the Sino-Japanese war in 1895, Japan accelerated the institutionalization of Okinawa. As part of the sovereign state of Japan, Okinawa was forced to assimilate. Top-down reforms took place in the administrative, economic, and education systems, despite protests and debates sparked over the desire to retain Ryukyuan indigenous social practices and languages.[8] The policy of assimilation had bloody consequences for Okinawa. Towards the end of World War II, Japan and the U.S. fought bitter battles on a number of Pacific islands, and the war reached Okinawa in the spring of 1945. Many residents were involved in the ground battle, giving up their land, livestock, family, and lives out of loyalty to Japan.[9] The battle

---

[4] Akamine, M. 2017. *The Ryukyu Kingdom: Cornerstone of East Asia*, Honolulu: University of Hawai'i Press: 32–57.

[5] Ibid.: 58–78.

[6] Smits, G. 1999. *Visions of Ryukyu: Identity and Ideology in Early-Modern Thought and Politics*, Honolulu: University of Hawai'i Press: 15–49.

[7] Ibid.: 144–149. There is a debate about Ryukyu as Japan's first colony followed by Taiwan and Korea. For a critical view regarding the question, see Christy, A. S. 1993. "The Making of Imperial Subjects in Okinawa," *Positions: Asia Critique* 1, no. 3: 607–39; Rabson, S. 1999. "Assimilation Policy in Okinawa: Promotion, Resistance, and 'Reconstruction,'" in Johnson, Chalmers (Ed.). *Okinawa: Cold War Island*, Cardiff, CA: Japan Policy Research Institute: 133–48; McCormack, G. & Norimatsu, S. O. 2012. *Resistant Islands: Okinawa Confronts Japan and the United States*, Lanham, MD: Rowman & Littlefield Publishers; Matsushima, Y. 2012. Ryukyu dokuritsu e no michi: Shokuminchi shugi ni aragau ryukyu nashonarizumu [*The Road to the Independence of Ryukyu: Ryukyu Nationalism Resisting Colonialism*] Kyoto: Hōritsu Bunkasha (In Japanese).

[8] Smits 1999: 148–9; Asato, S. Takara, K. Dana, M. Tomiyama, K. Nishizato, K. & Maehira, F. 2004. Okinawa-ken no rekishi [*The History of Okinawa Prefecture*]. Tokyo: Yamakawa Shuppansha: 256–294 (In Japanese). For example, there was a movement for recovering autonomy from Japan. Witnessing the signature of the Treaty of Shimonoseki as the settlement of the Sino-Japanese War in 1895, former members of the Ryukyu aristocratic class initiated an autonomy restoration movement. They originally aimed to recover their status in the new Okinawan society, facing a situation where people from the mainland occupied the high positions in the Okinawa administration and police force. Asato et al. 2004: 258–260. Another example is the so-called hōgen ronsō (Dialect Dispute), in which mainland Japanese folklore scholars criticized the Okinawan authorities' enforcement of standard Japanese and emphasized the protection of traditional culture. The prefectural education board counter-argued that standard Japanese was necessary for the Okinawan people to be able to communicate with the rest of the country and qualify as imperial subjects. A large debate ensued among Okinawan scholars as well. Asato et al. 2004: 293–294.

[9] Rabson 1999: 143–144.

became a war of attrition as the Japanese military intended to delay the possible U.S. invasion of mainland Japan as long as possible. Ultimately, approximately 120,000 Okinawans—about a quarter of the Okinawan population—were killed during the intense fighting on the ground.[10]

2.2 The U.S. Occupation Period

Mikio Higa, a scholar on Okinawan politics and the prefectural vice governor during the 1980s, noted that "post-war Okinawan political history was the history in which all the Okinawan people sought autonomy."[11] The end of the Okinawa battle was the beginning of the era of U.S. rule of the islands. As soon as U.S. forces seized the island, they began constructing bases for their future attack on the Japanese mainland. Meanwhile, residents and landowners were accommodated in civilian camps on the island. After the war, the San Francisco Peace Treaty was signed in 1951, granting the U.S. rights of administration, legislation, and jurisdiction over the Ryukyu Islands and other designated islands. Although a local administration, the Government of Ryukyu Islands (GRI), was established, it was subordinate to the U.S. Civil Administration of the Ryukyu Islands (USCAR). Under U.S. control, the people of Okinawa were given little freedom of speech, and taken off their farmland—expropriation characterized as a forced seizure by "bayonet and bulldozer"—and recruited to migrate to South America.[12] The people of Okinawa tried to oppose the despotic policies of the U.S., such as unreasonably low payments for acquisitioned land, lease, and compensation; further expansion of the bases; and increases in troop deployment, which were all facilitated by the demands posed by the U.S. anti-communist policy in Asia. However, the Chief Executive of the GRI, who was appointed by the USCAR, did not sincerely consider the Okinawan people's appeals, and the USCAR typically overruled the bills drafted by elected members of the Legislature.[13]

---

[10] McCormack and Norimatsu 2012: 17.

[11] Kono, Y. & Taira, Y. (Eds.) 2017. Taiwa Okinawa no sengo: Seiji rekishi shikō [*Dialogue Okinawa After War: Politics, History, and Thought*], Tokyo: Yoshida Shoten: 114 (In Japanese).

[12] As for the severity of the U.S. military's rule, see Aldous, C. 2003. "Achieving Reversion: Protest and Authority in Okinawa, 1952–70," *Modern Asian Studies*, Vol. 37, no 2: 485–508. Regarding post-war emigration from Okinawa to South America, see Amemiya, K. "The Bolivian Connection: U.S. Bases and Okinawan Emigration," In Chalmers (Ed.): 149–169; Selek, Y. 2003. "Migration and the Nation-State: Structural Explanations for Emigration from Okinawa." In Hook, G. D. & Siddle, R. (Eds.), *Japan and Okinawa: Structure and Subjectivity*, New York: RoutledgeCurzon: 74–92.

[13] Toriyama, A. 2015. Kichi to teikou [Bases and Resistance]. In Shimabukuro, J. & Abe, K. (Eds.), Okinawa ga tou nihon no anzen hoshō [*Revisiting Japan's Security from Okinawa*], Tokyo: Iwanami Shoten: 111–138 (In Japanese).

A highlight of the Okinawan resistance in the early period of U.S. occupation was the so-called "island-wide struggle" (*shima gurumi tōsō*), which took place after the U.S. announced a lump sum payment for land requisition in 1954. The GRI directly protested the purchases after interviews with landowners, asserting "four principles" that demanded the U.S. (1) refrain from purchasing the land, (2) pay appropriate lease fees and review payments annually, (3) appropriately compensate owners for loss of land, and (4) return abandoned land and avoid expropriating new land.[14] In June 1955, the Ryukyu government sent a delegation to Washington, D.C. for the first time to voice its opposition to the land purchase at the U.S. House of Representatives Committee of Armed Services.[15] Despite that, an American special group, which the committee assigned to visit Okinawa in October 1955, supported the predetermined plans involving a lump sum payment and further land requisition. This outcome disappointed and angered the people of Okinawa. It was shocking especially to the pro-U.S. Ryukyuan Democratic Party that believed Okinawa's autonomy and reversion to Japan could materialize only through cooperation with the U.S.[16] A nonpartisan protest movement formed and spread across the island.[17] Importantly, Okinawan politicians and schoolteachers linked their resistance to Japan's territorial integrity and fostered a mood among the Okinawan public in favor of the return of the islands to Japan.[18]

In addition, Okinawa began appealing to the government and people of Japan. When Okinawan politicians faced a standstill after failing to gain Washington's understanding, they shifted the target by asking Tokyo to negotiate with the U.S. on behalf of Okinawa.[19]

---

[14] Toriyama, A. 2013. Okinawa/ Kichi shakai no kigen to sōkoku [*Okinawa/ the Origin and Conflict in the Base Society*]. Tokyo: Keisō Shobō: 224–225 (In Japanese); Taira, Y. 2012. Sengo okinawa to beigun kichi: "Juyō" to "kyozetsu" no hazama de 1945–1972 [*Postwar Okinawa and U.S. Military Bases, 1945–1972*]. Tokyo: Hōsei Daigaku Shuppankyoku: 106 (In Japanese).

[15] Taira 2012: 105–115.

[16] Toriyama 2013: 241–243.

[17] The protest rally held simultaneously in each municipality on June 20, 1956, which was reportedly composed of 200,000 participants. Another rally took place 5 days later and about 150,000 gathered. At that time, the population of Okinawa was approximately 820,000. Taira 2012: 124–5.

[18] Tanji, M. 2006. *Myth, Protest and Struggle in Okinawa*, London: Routledge: 72–73.

[19] Taira 2012: 123–131; In addition, Toriyama points out that the people of Okinawa not only expected Japan to play a role on the issue but also had a doubtful view if Japan would take actions responsively. Toriyama 2013: 245–249.

2.3 Reversion to Japan and After

The movement for the return of Okinawa to Japan in the 1960s was characterized by deep emotions, hopes, and expectations. The movement was composed of people with a variety of motivations including nationalism, self-determination, democracy, fundamental human rights, and opposition towards war and nuclear weapons, and gave a strong impetus for Tokyo and Washington to agree on the reversion of Okinawa to Japan.[20] During the occupation period, the citizens of Okinawa had no means of interrupting U.S. military activities. They realized their lack of autonomy when facing insecurity and lawlessness after recurrent accidents and incidents caused by U.S. troops, as well as the deployment of nuclear weapons and bombing raids to Vietnam, giving rise to people's fear of involvement in an unwanted war.[21] Thus, many people in Okinawa seemed to believe that reversion would solve all the problems created by the U.S. presence, and that they would enjoy the protection of the Japanese constitution after the U.S. withdrawal.[22]

In 1967, Japan and the U.S. announced that they would begin negotiations on the return of the administrative rights of Okinawa to Japan within three years. However, the consequences of reversion to Japan differed from what was envisioned by the people of Okinawa. The gap between Okinawa's hopes of drastic reduction of U.S. bases and the reality that the bases remained after the return stemmed from a difference between Okinawa and Tokyo in the interpretation of the goal of reversion (i.e. "nuclear-free and homeland level status reversion" declared by Prime Minister Eisaku Sato in the Diet in March 1969).[23] During the negotiations with Washington, Tokyo sought to apply its constitution and the Japan-U.S. Security Treaty to Okinawa, while the U.S. tried to obtain a guarantee of free access and the introduction of nuclear weapons to the bases in Okinawa.[24] The remaining U.S. presence was a result of compromise.

In May 1972, Okinawa was returned to Japan as a prefecture. The main goal that Okinawa and Japan envisioned at the time was to reduce economic disparity between mainland Japan and Okinawa Prefecture. The former experienced rapid economic growth while the latter did not due to separation from the mainland caused by U.S. occupation. Tokyo created and invested in a 10-year development plan for Okinawa. It also established a compensation mechanism for Okinawan stakeholders such as

---

[20] Kawato, Y. 2015. *Protests Against U.S. Military Base Policy in Asia: Persuasion and Its Limits*, Stanford, CA: Stanford University Press: 53–64.

[21] Tanji 2006: 78–81.

[22] Ibid.: 104.

[23] The House of Councilors, Committee on Budget, March 10, 1969 (in Japanese). According to Nakajima, this was the first public statement in which Sato implied his plan to seek denuclearization of Okinawa after its reversion to Japan. Nakajima, T. 2012. Okinawa henkan to nichibei anpo taisei [*Reversion of Okinawa and the Japan-U.S. Security Arrangements*], Tokyo: Yūhikaku: 145–6. With regard to the negotiation on the issue between Okinawan and Japanese leaders, see Komatsu, H. "The Negotiating Process around 'Homeland Level Status' Reversion between Japan and Okinawa," *International Relations of the Asia-Pacific*, Vol. 18, No. 4: 1–28.

[24] Nakajima 2012.

owners of land used for U.S. bases and those who suffered from noise pollution and limited fishing.[25] The inflow of aid from the central government helped narrow the economic gap and enhance the standard of living. However, the Okinawan economy became dependent on the so-called 3Ks—*kichi* (bases), *kōkyō jigyō* (public works) and *kankō* (tourism).[26] This vertical relationship between the central and local governments eventually degraded Okinawa's autonomy by structuring its dependency on national subsidies and military bases.[27]

3 U.S. Bases in Okinawa in the Post–Cold War Era

Before examining the intention and action of Okinawa's search for autonomy, it is important to understand the international circumstances that surrounded Okinawa. This section reviews the U.S.'s view of its military presence in East Asia in the post-Cold War period. At the time, the U.S. had the rationale for maintaining a certain level of military forces in the Asia Pacific region. U.S. strategic reports, such as *A Strategic Framework for the Asian Pacific Rim*, published in 1990 and updated in 1992, indicate that the U.S. did not intend to drastically change the posture of its forces in Okinawa even after the Cold War ended. Moreover, Tokyo and Washington began coordinating the security arrangements required to meet certain regional challenges and, by the mid-1990s, came to a common understanding of the importance of U.S. forces stationed in the region.

3.1 Post–Cold War U.S. Presence in East Asia

Often described as a legacy of the Cold War, the Korean Peninsula was a U.S. security concern in the Asia Pacific in the early 1990s. Confronting North Korea, China, and the Soviet Union, the U.S. sustained its forward deployment in South Korea with around 38,000 to 40,000 personnel through the 1970s and 1980s.[28] Unlike in Western Europe, the dissolution of the Soviet Union had little impact on the Korean Peninsula.

[25] Cooley, A. and Marten, K. 2006 "Base Motives: The Political Economy of Okinawa's Antimilitarism," *Armed Forces & Society*, Vol. 32, No. 4: 571–575. Taira describes how the Defense Facilities Administration Agency arranged lease contracts before Okinawa reversion in response to the request made by private owners of land that the U.S. bases were located on. Taira 2012: 277–282.

[26] Miyamoto, K. 2010 "'Okinawa seisaku' no hyōka to tenbō" [The Assessment and Prospect of "Okinawa Policy"]. In Miyamoto, K. & Kawase, M. (Ed.), Okinawa Ron: Heiwa, Kankyō, Jichi no Shima e [*The Theory of Okinawa: Towards the islands of Peace, Environment, and Autonomy*]. Tokyo: Iwanami Shoten: 11–34 (In Japanese).

[27] Gavan, M. 2003. "Okinawa and the Structure of Dependence." In Hook & Siddle (Eds.): 93–113.

[28] Kane, T. 2016. *Global U.S. Troop Deployment, 1950–2005*, The Heritage Foundation, Retrieved from https://www.heritage.org/defense/report/global-us-troop-deployment-1950-2005

The U.S. was cautious about the uncertainty of North Korea's military modernization, especially its ambition to acquire nuclear capability.[29] Even though the U.S. considered a possible restructuring of the Army infantry division garrisoned in South Korea in 1990, the plan was postponed because of the North Korean attitude.[30]

The Bottom-Up Review, published by then-U.S. Secretary of Defense Les Aspin in October 1993, made it clear that the U.S. would maintain its strong commitment to the Asian region through its forward presence. Acknowledging that North Korea's military buildup could trigger an arms race in East Asia, the report stipulated the U.S. intention to retain a Marine Expeditionary Force in Okinawa.[31] Especially after the strained diplomacy between North Korea and international society led by U.S. efforts, it was expected that U.S. Marines would conduct an amphibious assault on North Korea from Okinawa.[32]

Another concern for the U.S. in the early 1990s was its sour relations with Japan. While the Clinton administration sought to adjust the trade imbalance with Japan, the bilateral relationship was drifting. Then-Prime Minister Morihiro Hosokawa appointed an advisory commission to review the 1976 National Defense Program Outline (NDPO), under which Japan contributed to regional peace and stability alongside the U.S., and searched for an alternative guideline for the new era. The report drafted by the commission recommended that Japan promote international and multilateral security cooperation through the United Nations and regional framework, such as the ASEAN (Association of Southeast Asian Nations) Regional Forum. Nonetheless, it clarified that the U.S.-centered alliance network would be irreplaceable and sustainable, even if U.S. relative power, particularly in the economic realm, would decline and international economic conflicts would be more likely.

However, American observers of Japanese politics interpreted the aforementioned recommendation as a hedge against the fading interests of the U.S. in the region,[33] reacting promptly to the reexamination process.[34] They were concerned that Japan's drift would have unfavorable consequences for the U.S. Reportedly, some within the Clinton Administration worried that if the U.S. hammered Japan on trade issues, the latter would stray from the alliance, which could threaten Asian countries with the thought of the possible remilitarization of Japan and endanger

[29] U.S. Department of Defense 1992. *A Strategic Framework for the Asian Pacific Rim: Report to Congress*.

[30] U.S. Department of Defense 1990. *A Strategic Framework for the Asian Pacific Rim: Report to Congress: Looking toward the 21st Century*; U.S. Department of Defense 1992.

[31] U.S. Department of Defense 1993. *Report on the Bottom-Up Review*.

[32] Collins, J. M. 1994. "Korean Crisis, 1994: Military Geography, Military Balance, Military Options." Congressional Research Service: 8.

[33] Cronin, P. M. & Green, M. J. 1994. *Redefining the U.S.-Japan Alliance: Tokyo's National Defense Program*, Washington DC: National Defense University, Institute for National Strategic Studies; Williams, D. 1995. "Rebuilding Military Ties to Tokyo: 'Nye Initiative' Launched to Address Post-Cold War Security Concerns," *The Washington Post*, February 19: A48.

[34] Ennis, P. 1995. "The 'Nye Initiative': Can It Save the U.S.-Japan Security Alliance?" *Tokyo Business Online*, June: 38–41; Asher, D. L. 1997. "A U.S.-Japan Alliance for the next Century," *Orbis*, Vol. 41, No. 3, June: 343–374.

vital U.S. bases in the country.[35] Their concerns yielded a new policy for Japan in the report, *The United States Security Strategy for the East Asia-Pacific Region*, known as the "Nye Initiative." The report published in February 1995 emphasized the deterrent role of U.S. military presence in the region. Based on the realist logic, it signaled that without U.S. presence Asian countries would step up military buildup and threaten neighbors, creating instability.[36] To prevent this situation and demonstrate commitment to security in the region, the report recommended maintaining the U.S. forward deployment of 100,000 personnel in the future. In the case of Japan, the U.S. intended to retain more than one Air Force wing, the Navy's Seventh Fleet activities, and a Marine Expeditionary Force in Okinawa for air and sea supremacy and rapid deployment.[37]

Subsequent close coordination between Tokyo and Washington for the betterment of bilateral relations resulted in changes of Japanese security policies. Japanese policymakers recognized that the outdated NDPO, which was drafted during the Cold War, needed to be adapted to the new situation.[38] The NDPO was revised in November 1995, stating that the U.S. continued to contribute to global peace and stability and that it was necessary to enhance the credibility of the Japan-U.S. security treaty.[39] The emphasis on bilateral security cooperation was consistent with the Joint Declaration on Security of April 1996. As per the declaration, the leaders of the two countries agreed that continued U.S. military presence was "essential for preserving peace and stability in the Asia-Pacific region," and shared the recognition that "smooth stationing" of the USFJ was the "core element" of the security arrangements.[40]

---

[35] Williams, "Rebuilding Military Ties to Tokyo."

[36] U.S. Department of Defense 1995. *United States Security Strategy for the East Asia-Pacific Region*: 9.

[37] Ibid.: 25.

[38] Mochizuki, M. M. 1995. *Japan: Domestic Change and Foreign Policy*. Santa Monica, CA: RAND: 71; Smith, S. A. 2003. "Beyond the "Post-Cold War"? U.S.-Japan Alliance and the Future of Asia-Pacific Security." In Ikenberry, J. G. & Inoguchi, T. (Eds.), *Reinventing the Alliance: U.S.-Japan Security Partnership in an Era of Change*. New York: Palgrave Macmillan: 121.

[39] Ministry of Foreign Affairs of Japan. National Defense Program Outline in and after FY 1996. December 1995. Retrieved from http://www.mofa.go.jp/policy/security/defense96/index.html; Akaha, T. 1998. "Beyond Self-Defense: Japan's Elusive Security Role under the New Guidelines for U.S.-Japan Defense Corporation." *The Pacific Review*, Vol. 11, No. 4: 465; Soeya, Y. 2005. "Japanese Security Policy in Transition: The Rise of International and Human Security." *Asia-Pacific Review*, Vol. 12, No. 1: 106–107.

[40] Ministry of Foreign Affairs of Japan 1996. Japan-U.S. Joint Declaration on Security: Alliance for the 21st Century. April 17. Retrieved from http://www.mofa.go.jp/region/n-america/us/security/security.html

3.2 A Crisis

On the evening of September 4th, 1995, three American servicemen kidnapped and raped a 12-year-old schoolgirl in Kin Town, Okinawa. This was not the first instance in which U.S. personnel committed crimes against local girls and women.[41] This incident reminded the local people of the brutal case in 1955, in which an American serviceman kidnapped, raped, and killed a 6-year-old girl.[42] One of the immediate results of the people of Okinawa's reaction to the current incident was their questioning of the Japan-U.S. Status of Forces Agreement (SOFA).[43] Since the SOFA allowed the U.S. military to detain the suspected servicemen until prosecution by Japanese authorities, local police were not able to take the three accused military personnel into custody.

The U.S. response to this incident was prompt. U.S. Ambassador to Japan Walter F. Mondale and Lieutenant General Richard B. Myers, the top U.S. commander in Japan, visited and apologized to Governor of Okinawa Masahide Ota on September 18th.[44] Furthermore, President William J. Clinton apologized on the radio, stating, "We will first of all make it clear that the United States deeply regrets the incident, that we do not condone any misconduct or any abuse of the Japanese people."[45] The U.S. Marine Corps in Okinawa held the "Day of Reflection" on October 5th and 18,000 marines attended lectures on the Okinawan culture, society, and customs as well as the SOFA.[46] Witnessing the sincere U.S. reactions, Ota concluded that the country understood the importance of restoring a stable presence in Okinawa to ensure better functioning of the Japan-U.S. security treaty.[47] The apologies and

---

[41] The advocacy group named Okinawa Women Act Against Military Violence compiled a list of crimes committed by U.S. servicemen against women from 1945 to the rape case in 1995. See http://www.jca.apc.org/~uratchan/gbmarines/henji/jyoseinohigai.html (in Japanese). However, it is noted that the official statistics indicate the tip of the iceberg, as sexual offenses tend to be unreported. Akibayashi, K. & Takazato, S. 2009. "Okinawa: Women's Struggle for Demilitarization." In Lutz, Catherine (Ed.), *The Bases of Empire: The Global Struggle Against U.S. Military Posts*, London: Pluto Press: 252.

[42] Miyagi, T. & Watanabe, T. 2016. Futenma Henoko Yugamerareta nijū nen [*Futenma Henoko the Distorted 20 Years*]. Tokyo: Shūeisha shinsyo: 21 (In Japanese).

[43] "Chiji tsuyoi fukaikan, chiikyōtei minaoshi yōkyu e" [Governor with strong displeasure, Requesting the SOFA revision]. Ryukyu shimpo, September 12, 1995, Morning edition: 1 (In Japanese).

[44] Jordan, M. 1995. "Rape of 12-Year-Old Fans Okinawans' Anger at U.S. Military Presence," *The Washington Post*, September 20: A15.

[45] "Regret From Clinton," *The Washington Post*, September 23, 1995: A15.

[46] "Okinawa beigun ga 'hansei no hi'" [U.S. Forces in Okinawa Held the 'Day of Reflection']. Mainichi Shinbun, October 6, 1995, Morning edition: 1 (In Japanese). Holding the 'Day of Reflection' was initiated by the Office of the Secretary Defense, not the Marine Corps. Funabashi, Y. 1999. *Alliance Adrift*. New York: Council on Foreign Relations: 301–302.

[47] Ota, M. 2000. Okinawa no ketsudan [*Okinawa's Decision*]. Tokyo: Asahi Shinbunsha: 175 (In Japanese).

reflections from the U.S. side showed that Washington took the case seriously and regarded the forward deployment in Okinawa as vital.

However, the repentant U.S. attitude did not ameliorate Okinawan antagonism towards the U.S. military. A protest rally took place across the prefecture on October 21st. The rally was significant in the post-reversion history of Okinawa in terms of size and inclusiveness. According to the executive committee of the rally, approximately 85,000 citizens participated. All political parties in the Okinawa Assembly joined, including the conservative Liberal Democratic Party that had supported coexistence with U.S. military bases. In addition, smaller protest gatherings took place in other islands of the prefecture and across Japan. The rally adopted a resolution composed of four appeals: (1) enforcing strict official discipline among U.S. military personnel and eradicating crimes by personnel and civilian employees of the U.S. military, (2) providing an immediate apology and complete compensation to the victim, (3) promptly reexamining the SOFA, and (4) accelerating consolidation and reduction of the bases.[48]

4 Okinawa's Ambition for Autonomy and its Impact Thereof

This section demonstrates that Okinawa tried to increase its autonomy over its own land, which had the potential to disrupt Japan's commitment to the alliance with the U.S. Contrary to Okinawa's efforts, the Japanese government legalized the centralization of procedures pertaining to national security matters in order to minimize the risk of offering land to the U.S. As a result, Okinawa lost its means of having its voice to be reflected in national policy.

4.1 Okinawa's Confrontation with the Alliance

Okinawan Governor Ota made a risky decision to escalate the base issues to the national level, whereas in the past they tended to be localized and overlooked since Okinawa's reversion to Japan. Before the infamous case of September 1995, Ota faced a difficult matter regarding the leasing contracts for the bases. The owners of the land appropriated for U.S. military use were not only the Japanese government, as was often the case with bases in mainland Japan, but also local municipalities and civilians. In the case of local owners disagreeing with the leases, the Special Measures Concerning Land for U.S. Armed Forces Law (Special Land Lease Law) provided procedures for using and expropriating the land offered to U.S. forces. This law specified that a governor was responsible for signing the lease documents on behalf of civilians and/or municipal owners who rejected a lease. Furthermore,

[48] "Kichi shukusyō shi Kyōtei minaose [Downsize the bases, Review the SOFA]." *Okinawa Times*, October 22, 1995, Morning edition: 1 (In Japanese).

the lease was valid for five years. The law considered a governor's proxy signature as an "assigned function," as previous Okinawan governors had signed such documents. Ota was one such governor. He went ahead with the proxy signature during his first term in 1991. He had considered scrapping the procedure since the gubernatorial election campaign, but did not. This was because he expected the central government to establish a new institution to solve base issues in Okinawa and prioritized a good relationship with the central government for benefit of the upcoming 10-year development plans for Okinawa.[49] However, as the second round of permissions for the leases was approaching, he debated whether or not to follow the established procedures.[50]

At the Prefectural Assembly regular meeting on September 28th, 1995, Governor Ota announced his decision not to sign the land-lease documents and explained his reasons.[51] First, he noted Okinawa's history with the U.S. military, which established its bases by expropriating Okinawan land in an oppressive manner. As much as 75% of the U.S. military facilities in Japan were concentrated in Okinawa Prefecture, which accounts for 0.6% of Japanese land area, and the large bases hindered the prefecture from developing. Second, he argued that U.S. military presence negatively impacted the people of Okinawa—their training activities caused environmental damage, noise pollution, and aircraft accidents. Moreover, there were numerous criminal offenses committed by American personnel, of which the incident involving the schoolgirl was just one example. Third, he was concerned that the U.S. presence in Okinawa would endure in the future. He expected that the end of the Cold War would foster the consolidation and reduction of the number of bases in Okinawa as well as in other parts of the world. However, the strategic report for East Asia specified that the U.S. would sustain the 100,000 personnel deployed to the region.[52] Furthermore, Japan and the U.S. evaluated the security treaty and reviewed the operations of U.S. bases in Japan from global perspectives. In this regard, Ota was deeply concerned that the bases in Okinawa would be reinforced and perpetuated, which contrasted with his vision for the twenty-first century.

In addition to these reasons, certain other factors also contributed to the governor's decision; for example, a meeting with then-Minister of Foreign Affairs Yohei Kono immediately after the assault case. Listening to Ota's request to revise the SOFA, Kono responded that there was a gap in the appeal, which confirmed that

[49] Ota 2000: 123

[50] Within the prefectural government, there was a debate on the signature process. Then-Vice Governor Nakaima advised Ota that it was just an administrative procedure, not a yes-no question. The National Graduate Institute for Policy Studies (GRIPS) 2005. Yoshimoto Masanori (Moto okinawaken fukuchiji) Ōraru hisutorī [*Yoshimoto Masanori (Former Vice Governor of Okinawa) Oral History*] Tokyo: National Graduate Institute for Policy Studies: 44 (In Japanese).

[51] Okinawa Prefectural Assembly, Regular Meeting on September 28, 1995 (In Japanese).

[52] U.S. Department of Defense 1995. *United States Security Strategy for the East Asia-Pacific Region.*

Tokyo did not share the same anger and discontent as Okinawa.[53] Moreover, according to then-Vice Governor Masanori Yoshimoto, Ota had felt powerless when visiting Washington to request that the number of bases be reduced. The U.S. government was reluctant to accept his proposal, stating that, "the Japanese government has not mentioned the consolidation and reduction of bases in Okinawa. We would consider such action if proposed."[54] With these increasing concerns in mind, Ota considered the administrative procedure of proxy signature an opportunity to make the central government listen to the local voice. The purpose of his decision to do so was to encourage the Prime Minister of Japan to refuse the lease agreements and negotiate with the U.S. on the base issues as a sovereign state.[55]

At the same time, the Ota prefectural administration tried to develop an independent economic structure. When the Okinawa Promotion Plan, which was the third 10-year development plan since Okinawa's reversion to Japan, was launched in 1992, Okinawa independently elaborated a development concept aimed at self-sustaining development by employing regional characteristics. Aware of ongoing and future globalization, the concept, which was later named the Okinawa International City Plan, aimed to develop Okinawa as a "bridge to the world" through economic, industrial, and cultural exchange; peace diplomacy; and international technical cooperation.[56] The Okinawa government released the promotion plan in March 1996 along with the Base Reversion Action Program, which demanded the gradual return of the land of all U.S. bases over a period of 15 years in order to redevelop the area. Behind these plans for Okinawa was the assumption by prefectural officials that the end of the Cold War would lead to the downsizing of U.S. forces in Okinawa and that, likely, the central government would not continue implementing special measures for development.[57]

---

[53] Miyagi, E. 1996. "Redressing the Okinawan Base Problem," *Japan Quarterly*, Vol. 43, No. 1: 30; Eldridge, R. D. 1997. "The 1996 Okinawa Referendum on U.S. Base Reductions: One Question, Several Answers," *Asian Survey*, Vol. 37, No. 10: 882. On the other hand, Kono thought it would be more important and effective to change the procedures of handing over the suspects than revising the principle. Funabashi 1999: 305.

[54] GRIPS 2005: 75.

[55] Okinawa Prefectural Assembly, Regular Meeting on December 7, 1995 (In Japanese).

[56] Okinawa-ken 1994. Kokusai toshi keisei seibi kōsō chōsa II hōkokusyo [*Report on the International City Formation and Development Concept II*]. Naha: Okinawa-ken (In Japanese). According to the consignee's report, an international city is defined as one with special regional characteristics or functions that are attractive and recognized globally. Okinawa-ken 1993. Okinawa hontō chūnanbu toshiken kokusai toshi keisei seibi kōsō chōsa hōkokusyo [*Report on the International City Formation and Development Concept for the Middle and Southern Okinawa Urban Areas*]. Naha: Okinawa-ken: 73–74 (In Japanese).

[57] Shimabukuro, J. 2015. Okinawa jiritsu no kōsō to higashi ajia no kōzōtenkan [The Concept of Okinawan Autonomy and the Structural Transition of East Asia]. In Shimabukuro & Abe: 295. Ota supposed that the U.S. force reduction would be possible, observing the discussion on the withdrawal of U.S. Marine Corps in Washington. See GRIPS 2003. Ōraru hisutorī Ota Masahide (Moto okinawa-ken chiji) [*Oral History Ota Masahide (Former Governor of Okinawa)*, Tokyo: National Graduate Institute for Policy Studies: 91 (In Japanese). Vice Governor Yoshimoto also stated it was assumed that U.S. bases could be reduced due to the prospective improvement of situations of the Korean Peninsula and Taiwan. GRIPS 2005: 58 (In Japanese).

Note that the locally generated plans included Okinawa's ambition for autonomy, although it was stated that Okinawa's development would contribute to the future progress of Japan as a whole. The Okinawa International City Plan recommended an All-Okinawa Free Trade Zone, tax reduction and exemption, and simplifying immigration procedures, among other things, implying a partial delegation of the central government's authority to the governor.[58] Furthermore, some scholars analyzed that Okinawa sought not only to apply the "One Country Two Systems" approach, but also to cultivate a foundation for independence.[59]

Besides the governor and the local government, the quest for autonomy was expressed at the citizen level. After the historical protest rally in October 1995, a labor union association became the driving force in preparing a prefectural plebiscite to question voters about the revision of the SOFA and the consolidation and reduction of U.S. bases. The purpose of the referendum was not to settle the base issues, but to influence the political decision process in the international arena.[60] The plebiscite was held on September 8th, 1996, finding that 89% of voters agreed with the SOFA revision and the base reduction, even though voter turnout was less than 60%. Despite criticism of the plebiscite,[61] the referendum took place for the first time in Okinawa to demonstrate public support for the governor to confront Tokyo and residents' will to change the status quo of the society.

4.2 *The Alliance's Response*

A sense of crisis drove Japanese and American alliance managers to deal with base issues in Okinawa immediately. A risk for the Japanese government was that it would lose a legal foundation from which to offer land for U.S. bases if the governor

[58] Nakachi, H. 2005. Kokusaitoshi keisei kōsō to dōshūsei [A Grand Design for International City and the Regional System]. *Ryudai Law Review,* Vol. 73: 189–207 (In Japanese). Note that the proposed All-Okinawa Free Trade Zone stirred much controversy within Okinawa. Hook, G. D. 2003. "Responding to Globalization: Okinawa's Free Trade Zone in Microregional Context," In Hook & Siddle: 39–54; Sasaki, M. 1999. "Sustainable Development in Okinawa for the 21st Century," In Chalmers: 247–260.

[59] Shinohara, Akira. "Nihon no shisutemu tou okinawa no 'dokuritsu' kōsō [Okinawa's 'Independence' Concept Questioning Japanese System]". Asahi Shinbun, October 8, 1996, Evening edition, p.15 (In Japanese); Gabe, M. 1997. "Ikkoku nisheido ga dokuritsu maneku [the One Country Two Systems will cause independence]". Asahi Shinbun, June 23, Morning edition: 4 (In Japanese). Meanwhile, an expert on free trade claimed that Okinawa's free trade zone plan would require the status of a self-governing dominion rather than independence. Otani, R. Y. 1997. "Furīpōto wo okinawa ni [Bring free port to Okinawa]". Ryukyu shimpo, June 13, Morning edition: 3 (In Japanese).

[60] Shimabukuro, J. 1997. "Okinawa kenmin tōhyō ni okeru seiji katei [Political Process in the Okinawa Prefectural Plebiscite]," *Toshi Mondai,* Vol. 89, No. 2: 23 (In Japanese).

[61] Eldridge 1997: 888–898; Egami, T. 1998. "Okinawa no kenmin tōhyō [Okinawan Prefectural Referendum]," *Review of Policy Science and International Relations*, Vol. 1: 15–16 (In Japanese).

did not agree on the leases by the expiry date. Three weeks after the incident was revealed, Foreign Minister Yohei Kono met U.S. Ambassador to Japan Walter Mondale, seeking cooperation to cope with the consolidation and reduction of U.S. bases in Okinawa.[62] In November, the two countries formed a special organization, the Special Action Committee on Okinawa (SACO), to discuss reducing the "burden" on the people of Okinawa. Then-Secretary of Defense William Perry suggested that the name of the committee include 'action' because he believed that the meeting should result in actions and visible outcomes, not merely discussions and recommendations.[63]

To mitigate the impacts of the U.S. military presence on Okinawa, after about a year of intensive consultations between Japan and the U.S., SACO proposed various solutions in its final report in December 1996. These ranged from the return of superabundant or auxiliary facilities to the relocation of artillery firing training and a transportation unit to mainland Japan. The featured recommendation was the return of the Marine Corps Air Station Futenma. The airbase in the middle of a residential city of Ginowan, located about 12 kilometers north of Naha, was one of the bases the Okinawan government had appealed to Tokyo and Washington to return since the 1980s. However, it was an important base for the Marine Corps in terms of providing airlift for ground troops, which enabled them to respond quickly to contingencies in the Asia-Pacific region. In short, driven by imperatives to ease the impacts and secure a better environment for sustaining the bases, SACO recommended that the U.S. forces in Okinawa, especially the Marine Corps, arrange their activities off the island as much as possible.

However, the SACO proposal did not present radical or fundamental changes to the stationing U.S. forces. Most of the proposed base returns required alternative sites in Okinawa. The Futenma base was not an exception—the final report recommended that a replacement should be a sea-based facility off the east coast of the Okinawan main island. Governor Ota was concerned over the fact that the municipalities where the facilities were to be relocated had expressed opposition.[64] Furthermore, the revision of the SOFA, for which Okinawa had strongly appealed, was not included in the proposal, although the two governments confirmed detailed operational measures.[65] From Okinawa's viewpoint, the SACO proposals were less than satisfactory without drastic measures to redress the problems they had faced for a long time.

On the other hand, the U.S. had a different purpose for SACO from the onset. A briefing memo prepared by Curt Campbell, then-Deputy Assistant Secretary of Defense for Asia and the Pacific, in April 1996 before the launch of the SACO

---

[62] "Okinawa kichi shukushō ni saidai kyōryoku [Maximum Cooperation for the Base Reduction in Okinawa]," Yomiuri Shinbun, October 3, 1995, Evening edition: 1 (In Japanese).

[63] Funabashi 1999: 308–309.

[64] "Kennai isetsu zentei ni hanpatsu ga aru [Opposition to the Prerequisite of the Relocations within Okinawa Prefecture]," *Mainichi Shinbun*, December 3, 1996, Morning edition: 3 (In Japanese).

[65] Japan Ministry of Foreign Affairs 1996. "The SACO Final Report," December 2. Retrieved from http://www.mofa.go.jp/region/n-america/us/security/96saco1.html

interim report in Tokyo, revealed that the purpose of SACO was to sustain training and operational readiness, not to modify U.S. force posture or the deployment of 47,000 personnel in Japan.[66] In addition, then-Secretary of Defense Perry later confessed that "we [the U.S.] never considered giving up the [Futenma] base. We always considered what we could do to keep the capability that we felt was vital," because the U.S. recognized that the primary role of the base was to aid in rapid reinforcement of its forces in South Korea in the case of an offensive by North Korea.[67] The bilateral special committee disbanded after these recommendations.

4.3 Tokyo's Response

The fact that Ota raised the issue of land leases for U.S. bases had impact on some alliance managers in the Japanese government. The government decided to accelerate the consolidation and integration of bases based on its awareness that the core of the Japan-U.S. security relationship could be damaged, because anti-base sentiment was emerging in Okinawa and political factions had voiced critical reviews of the bases and the SOFA.[68] Especially in the Defense Agency (transitioned to the Ministry of Defense in 2007), which had long coped with locals on base-related matters, the view was growing that the current problem might not be resolved unless the government pushed to reduce or return the U.S. bases.[69] The government was concerned that it could severely undermine the credibility of Japan if it offered the U.S. military land with no title and made its presence illegal.[70] Ota's challenge forced the central government to reconsider how to maintain the basis of the alliance.[71]

Not only did Tokyo establish a taskforce with Washington, as mentioned in the previous section, but also a consultation body with Okinawa. Ota requested setting up a meeting wherein Tokyo and Okinawa could directly discuss U.S. base-related problems and requests from Okinawa. This was a remarkable initiative at the time, because the central and prefectural governments rarely entered into such con-

[66] U.S. Department of Defense 1996. "Secretary of Defense Perry Visit to Tokyo, 14–15 April 1996" [Background Book]. In Digital National Security Archive, DNSA Collection, Japan and the United States: Diplomatic, Security, and Economic Relations, Part III, 1961–2000. Retrieved from https://search.proquest.com/docview/1679142620?accountid=14891

[67] Tsuruya, K. 2017. Perī no kokuhaku: Moto kokubōchōkan Okinawa eno tabi [Perry's Confession: The Former Defense Secretary's Trip to Okinawa] [Television broadcast]. Tokyo: NHK, November 18 (In Japanese).

[68] "Chiikyōtei to beigun kichi mondai [The SOFA and U.S. Military Base Issues]," *Nihon Keizai Shinbun*, October 4, 1995: 2 (In Japanese).

[69] Akiyama, M. 2002. Nichibei no senryakutaiwa ga hajimatta: Anpo saiteigi no butaiura [*Start of Japan-U.S. Strategic Dialogue*], Tokyo: Aki Shobō: 80 (In Japanese).

[70] Ibid.: 190.

[71] Miyagi 1996.

sultation on an equal footing.[72] Then-Prime Minister Tomiichi Murayama agreed to establish the special committee for Tokyo and Okinawa, because this seemed to be the natural course of action: "Okinawa has been fulfilling its duty and made sacrifices for Japan as a whole."[73] However, the purpose of the consultation initiatives, including SACO, was reportedly to ensure that the prime minister would sign the lease by proxy.[74]

Next, the central government entrusted the issue to the judiciary, which ultimately withheld Okinawa's autonomy over the land in question. The prime minister initiated a court battle in December 1995 by filing a suit against the governor, because the latter's refusal was considered insubordination to the Local Autonomy Law. At the first trial, Okinawa argued that forcibly using citizens' land for U.S. bases was an infringement of the property rights guaranteed by the Japanese Constitution, as these bases did not serve the public welfare. It further claimed that if the assigned function infringed citizens' human rights and hindered their welfare, a governor could refuse the function, which should be a matter of local autonomy.[75] In March 1996, the first judge favored the central government, considering that the governor's proxy signing was to ensure the smooth conduct of land acquisition procedures. The Okinawa government submitted a final appeal the next month. However, in August 1996, the Supreme Court confirmed that it was not the governor but the prime minister who was qualified to judge whether the land could be offered to U.S. forces.[76] It concluded that the governor's refusal to sign by proxy would hamper the Japanese government from fulfilling its duty to the Japan-U.S. security treaty.[77]

The impasse over the land lease issue between the central and local governments ended after the judgment by the Supreme Court. In September 1996, Ota agreed that the land lease process should proceed to the next stage of public notice and inspection. While his supporters in academia advised the governor against participating in the procedures, he expressed his apprehension about the possible establishment of special legislation,[78] which may have meant shutting the window of opportunity for

[72] Ota 2000: 182–187.

[73] Ibid.: 186.

[74] "Okinawa kichi kyōgi kaishi [Discussion on Okinawa Bases Started]," Yomiuri Shinbun, November 26, 1995, Morning edition: 2 (In Japanese).

[75] Takara, T. 2005. "Dairishomei soshō: heiwa wo uttaeta saiban [The Lawsuit of Substitute Signature: The Trial of Appealing Peace]," In Ishihara, Masaie, Nakachi, Hiroshi & Lummis, Douglas C. (Eds.), Okinawa wo heiwagaku suru! [*Doing Peace Studies in Okinawa!*] Kyoto: Hōritsu Bunkasha: 92 (In Japanese).

[76] Supreme Court of Japan 1996. "Judgment upon Case of Constitutionality of the Forced Leasing of Land for U.S. Bases in Okinawa Prefecture." Retrieved from http://www.courts.go.jp/app/hanrei_en/detail?id=268; Hirota, Kyōichi. "Okinawa beigun kichi mondai no shinjitsu to tenbō [Okinawa's U.S. Military Base Issue: Reality and Prospects]," *Journal of National Defense*, Vol. 24, No. 3: 65–82 (In Japanese).

[77] "Okinawa no uttae shirizokeru [The Court Rejected Okinawa's Appeal]," *Asahi Shinbun*, August 29, 1996, Morning edition: 1 (In Japanese).

[78] Shiohira, Y. 1997. "Okinawa kenmin tōhyō, sono go [The Okinawa Prefectural Referendum and After]," *Toshi Mondai*, Vol. 89, No. 2: 76 (In Japanese).

Okinawa to reflect its intention to the center. In fact, before his decision to yield, Ota secretly requested for then-Prime Minister Ryutaro Hashimoto not to revise the Special Land Lease Law.[79]

However, contrary to Ota's hopes, Tokyo sought to mitigate the impact of peripheral autonomy by concentrating authority over national security matters. After the court case, there was a possibility that the leases for another 12 facilities would expire by May 1997. Tokyo was unwilling to become involved in the revision of the Special Land Lease Law, which could intensify the confrontation with Okinawa.[80] Nevertheless, it approved a partial revision that permitted the provision of land under examination by the Expropriation Commission. Hashimoto explained that the Japanese national interest was to maintain the Japan-U.S. security arrangement; thus, it was the Japanese responsibility to provide U.S. forces with facilities and areas.[81] A month before expiration, both the House of Representatives and the House Councilors passed the revision bill that enabled U.S. forces to continue using the land before completing the necessary lease procedures.

The central government further solidified lease procedures to ensure the stable provision of land to the ally. The Committee for the Promotion of Decentralization, an advisory body launched in July 1995, convened and presented five recommendations. The third recommendation, submitted in September 1997, focused on the Special Land Lease Law.[82] The committee recommendation stated that this law was considered part of the state's land provision in accordance with the obligation to the Japan-U.S. Security Treaty. To clarify the division of labor between the state and municipalities, the committee concluded that the central government should be responsible for national and international obligations, while governors and mayors should represent their municipalities. Heeding this advice, the Japanese government began preparing to revise the Special Land Lease Law. It was included in the package regarding the Law on Decentralization of Government Power and passed by an overwhelming majority in July 1999. Then-Okinawan Governor Keiichi Inamine, who assumed office in 1998, requested that the mechanism be maintained so that the relevant municipalities and prefectures could become involved in procedures to lease the bases. He voiced his concerns that the people's intentions would not be reflected.[83]

[79] Ota 2000: 240.

[80] "Kinkyū shiyō nigatsu nai ni moushitate [The Japanese Government Will Make a Statement of Emergency Use in February]," Asahi Shinbun, February 9, 1997, Morning edition: 1 (In Japanese).

[81] The House of Councilors, Committee on Budget, March 6, 1997 (In Japanese).

[82] The Committee for the Promotion of Decentralization 1997. Dai san-ji kankoku: Bunken gata shakai no sōzō [*The Third Recommendation: The Creation of Decentralized Society*]. Tokyo: Cabinet Office, Government of Japan, September 2, Retrieved fromhttp://warp.ndl.go.jp/info:ndljp/pid/8313852/www8.cao.go.jp/bunken/bunken-iinkai/3ji/ (In Japanese).

[83] "'Kuni ga dairi shomei' ni okinawa hanpatsu [Okinawa Opposes to a 'Substitute Signature by the National Government']," Asahi Shinbun, June 12, 1999, Morning edition: 3 (In Japanese).

Along with revising the law, Tokyo utilized economic incentives to appease discontent with the heavy "burden" of U.S. bases among the people of Okinawa.[84] At the prime minister-governor meeting in September 1996, Hashimoto announced that the central government would create a new joint panel, the Okinawa Policy Council, in which the governor and relevant ministers would discuss base-related issues as well as development measures. Furthermore, he committed to allocating a special fund of five billion yen to promote an independent local economy, boost employment, and enhance people's quality of life.[85] Clearly, Ota faced a dilemma. He could have persisted in objecting to the U.S. bases, as his supporters advised, but Tokyo's cooperation was much needed for the Okinawan economy and development. Ota chose to accept the deal because it would enable the local government to carry out the regional promotion projects it had elaborated.

5 Conclusion

In conclusion, I will answer the questions delineated in this chapter. First, Okinawa sought to increase its autonomy by emphasizing the historically accumulated negative impacts of U.S. forces on the lives of those who inhabit the island prefecture at opportunistic times following incidents such as the rape case of 1995. After the end of the Cold War, Okinawa expected a peace dividend—that is, drastically reduced U.S. military presence on the islands. However, the U.S. advanced a security initiative for the Asia-Pacific region to sustain the deployment of 100,000 personnel. Concerned about continued concentration of U.S. troops and repeated crimes by U.S. servicemen, Governor Ota considered the land lease proxy signature, which had generally been conducted as a purely administrative procedure, as an opportunity to express the intent not to continue providing land for U.S. bases in the future. In so doing, he confronted Tokyo, asking decision-makers to reconsider the Okinawa situation and the Japan-U.S. security treaty arrangement. He gained the support of various groups and citizens in and outside of the prefecture. A non-partisan rally was attended by the largest number of participants in the post-reversion history. Also, the first prefectural referendum confirmed that a majority of voters was in favor of revising the SOFA and downsizing the U.S. bases. These were manifestations of the people's voices through direct democracy. As a result, Okinawa raised itself to a position that in effect exceeded its status as a prefecture when Tokyo set up a council where the Okinawan governor was able to negotiate with relevant

[84] For detailed analysis of Tokyo's economic incentive as compensation and persuasion, see George Mulgan, A. 2000. "Managing the U.S. Base Issue in Okinawa: A Test for Japanese Democracy." *Japanese Studies*, Vol. 20, No. 2: 166–171; Calder, K. E. 2007. *Embattled Garrisons: Comparative Base Politics and American Globalism*. Princeton: Princeton University Press: 129–139.

[85] "Okinawa kichi mondai ni tsuite no hashimoto shushō danwa [Prime Minister Hashimoto's Comment on U.S. Base Issues in Okinawa]," *Mainichi Shinbun*, September 11, 1996, Morning edition: 2 (In Japanese).

ministers about base issues on an equal footing. That, however, was not a position that would guarantee autonomy in trilateral consultations with Tokyo and Washington.

Second, Okinawa's quest for autonomy conflicted with Japanese national security interests because of the Japanese government's responsibility to provide the U.S. basing access. Ota's refusal of the proxy signature was a concern to Tokyo because the lack of a legal foundation to provide land for the U.S. bases would undermine Japan's credibility regarding the security treaty. Underlying the concern were the close bilateral arrangements for redefining Japan-U.S. security relations. It was a mutually confirmed item on the national agenda that the U.S. would not change its force structure in the Asia-Pacific region, including USFJ. Furthermore, the U.S. did not intend to meet Okinawa's demands to reduce its military presence if it required that the force posture be modified. However, the sub-state actor Okinawa managed to influence bilateral security arrangements to the extent that Japan and the U.S. agreed on returning the Futenma air station and other facilities as well as relocating several training activities out of the prefecture.

Third, Okinawa's push for autonomy was restrained by the judicial and legislative power of the center. Okinawa's challenge to the existing land lease procedures resulted in a judgment against it and a clear separation of administrative responsibilities between the central and local governments. After the revision of the Special Land Lease Law, Okinawa lost its means to resist the power of the center. On the other hand, Tokyo was able to offer the U.S. stable basing access without hindrance. In addition, Tokyo dangled economic incentives to appease anti-base emotions. Prime Minister Hashimoto pledged to provide Okinawa with financial and economic assistance to advance the Okinawa-led development plan. To remain in good standing with Tokyo, Ota was forced to relinquish his confrontational attitude regarding the U.S. bases.

References

Akaha, T. 1998. Beyond self-defense: Japan's elusive security role under the new guidelines for U.S.-Japan defense corporation. *The Pacific Review* 11 (4): 461–483.

Akamine, M. 2017. *The Ryukyu kingdom: Cornerstone of East Asia*. Honolulu: University of Hawai'i Press.

Akibayashi, K. and S. Takazato. 2009. Okinawa: Women's struggle for demilitarization. In *The bases of empire: The global struggle against U.S. military posts*, ed. Lutz, C., 243–269. London: Pluto Press.

Akiyama, M. 2002. *Nichibei no senryakutaiwa ga hajimatta: Anpo saiteigi no butaiura [Start of Japan-U.S. strategic dialogue]*. Tokyo: Aki Shobo.

Aldous, C. 2003. Achieving reversion: Protest and authority in Okinawa, 1952–70. *Modern Asian Studies* 37 (2): 485–508.

Asato, S., K. Takara, M. Dana, K. Tomiyama, K. Nishizato and F. Maehira. 2004. *Okinawa-ken no rekishi [The history of Okinawa prefecture]*. Tokyo: Yamakawa Shuppansha.

Asher, D.L. 1997. A U.S.-Japan alliance for the next century. *Orbis* 41 (3): 343–374.

Calder, K.E. 2007. *Embattled garrisons: Comparative base politics and American globalism*. Princeton: Princeton University Press.

Christy, A.S. 1993. The making of imperial subjects in Okinawa. *Positions: Asia Critique* 1 (3): 607–639.

Collins, J.M. 1994. *Korean crisis, 1994: Military geography, military balance, military options*. Washington, DC: Congressional Research Service.

Cooley, A. 2008. *Base politics: Democratic change and the U.S. military overseas*. Ithaca: Cornell University Press.

Cooley, A. and K. Marten. 2006. Base motives: The political economy of Okinawa's antimilitarism. *Armed Forces & Society* 32 (4): 566–583.

Cronin, P. M. and M. J. Green. 1994. *Redefining the U.S.-Japan alliance: Tokyo's national defense program*. McNair Paper 31. Washington, DC: National Defense University, Institute for National Strategic Studies.

Egami, T. 1998. Okinawa no kenmin tōhyō [Okinawan prefectural referendum]. *Review of Policy Science and International Relations* 1: 1–23.

Eldridge, R.D. 1997. The 1996 Okinawa referendum on U.S. base reductions: One question, several answers. *Asian Survey* 37 (10): 879–904.

Ennis, P. 1995. The "Nye initiative": Can it save the U.S.-Japan security Alliance? *Tokyo Business Online*: 38–41.

Funabashi, Y. 1999. *Alliance adrift*. New York: Council on Foreign Relations.

George Mulgan, A. 2000. Managing the U.S. base issue in Okinawa: A test for Japanese democracy. *Japanese Studies* 20 (2): 159–177.

Hirota, K. 1996. Okinawa beigun kichi mondai no shinjitsu to tenbō [Okinawa's U.S. military base issue: Reality and prospects]. *Journal of National Defense* 24 (3): 65–82.

Hook, G.D. 2003. Responding to globalization: Okinawa's free trade zone in microregional context. In *Japan and Okinawa: Structure and subjectivity*, ed. Hook, G.D. and R. Siddle, 39–54. New York: RoutledgeCurzon.

Japan Ministry of Defense. 2017. *Defense of Japan 2017*. Retrieved from http://www.mod.go.jp/e/publ/w_paper/2017.html.

Johnson, C. 1999. *Okinawa: Cold war island*. Cardiff: Japan Policy Research Institute.

Kane, T. 2006. *Global U.S. Troop deployment, 1950–2005*. Retrieved from https://www.heritage.org/defense/report/global-us-troop-deployment-1950-2005.

Kawato, Y. 2015. Protests against U.S. military base policy in Asia: Persuasion and its limits. Stanford: Stanford University Press.

Komatsu, H. 2017. The negotiating process around 'homeland level status' reversion between Japan and Okinawa. *International Relations of the Asia-Pacific* 18 (1): 1–28.

Matsushima, Y. 2012. *Ryukyu dokuritsu e no michi: Shokuminchishugi ni aragau ryukyu nashonarizumu* [The path to the independence of Ryukyu: Ryukyu nationalism resisting colonialism]. Kyoto: Hōritsu Bunkasha.

McCormack, G. 2003. Okinawa and the structure of dependence. In *Japan and Okinawa: Structure and subjectivity*, ed. Hook, G.D. and R. Siddle, 93–113. New York: RoutledgeCurzon.

McCormack, G. and S.O. Norimatsu. 2012. *Resistant Islands: Okinawa confronts Japan and the United States*. Lanham: Rowman & Littlefield Publishers.

Ministry of Foreign Affairs of Japan. 1996a (April 17). Japan-U.S. joint declaration on security: Alliance for the 21 Century. Retrieved from http://www.mofa.go.jp/region/n-america/us/security/security.html.

———. 1996b (December 2). *The SACO final report*. Retrieved from http://www.mofa.go.jp/region/n-america/us/security/96saco1.html.

Miyagi, E. 1996. Redressing the Okinawan base problem. *Japan Quarterly* 43 (1): 27–32.

Miyagi, T. and T. Watanabe. 2016. *Futenma Henoko yugamerareta nijū nen [Futenma Henoko the distorted 20 years]*. Tokyo: Shūeisha Shinsyo.

Miyamoto, K. 2010. 'Okinawa seisaku' no hyōka to tenbō [the assessment and prospect of Okinawa policy]. In *Okinawa Ron: Heiwa, kankyō, jichi no Shima e [The theory of Okinawa:*

Towards the islands of peace, environment, and autonomy], ed. Miyamoto, K. and M. Kawase, 11–34. Tokyo: Iwanami Shoten.
Mochizuki, M.M. 1995. *Japan: Domestic change and foreign policy*. Santa Monica: RAND.
Nakachi, H. 2005. Kokusaitoshi keisei kōsō to dōshusei [A grand design for international city and the regional system]. *Ryudai Law Review* 73: 189–207.
Nakajima, T. 2012. *Okinawa henkan to nichibei anpo taisei [Reversion of Okinawa and the Japan--U.S. security arrangements]*. Tokyo: Yūhikaku.
National Graduate Institute for Policy Studies (GRIPS). 2003. *Ōraru hisutorī Ota Masahide (moto okinawa-ken chiji) [Oral history Ota Masahide (former governor of Okinawa)]*. Tokyo: National Graduate Institute for Policy Studies.
———. 2005. *Yoshimoto Masanori (moto okinawa-ken fukuchiji) Ōraru hisutorī [Yoshimoto Masanori (former vice governor of Okinawa) oral history]*. Tokyo: National Graduate Institute for Policy Studies.
Okinawa Prefectural Government Washington DC Office. n.d. *U.S. military base issues in Okinawa*. Retrieved from http://dcoffice.org/basedata#p3.
Okinawa-ken. 1993. *Okinawa hontō chūnanbu toshiken kokusai toshi keisei seibi kōsō chōsa hōkokusyo* [Report on the international city formation and development concept for the middle and Southern Okinawa urban areas]. Naha: Okinawa-ken.
———. 1994. *Kokusai toshi keisei seibi kōsō chōsa II hōkokusyo* [Report on the international city formation and development concept II]. Naha: Okinawa-ken.
———. 2017. *Okinawa no beigun oyobi jieitai kichi* [Bases of U.S. military forces and the Japanese self-defense forces in Okinawa]. Naha: Okinawa-ken.
Ota, M. 2000. *Okinawa no ketsudan* [Okinawa's decision]. Tokyo: Asahi Shinbunsha.
Rabson, S. 1999. Assimilation policy in Okinawa: Promotion, resistance, and "reconstruction". In *Okinawa: Cold war island*, ed. Johnson, C., 133–148. Cardiff: Japan Policy Research Institute.
Sasaki, M. 1999. Sustainable development in Okinawa for the 21st Century. In *Okinawa: Cold war island*, ed. Johnson, C., 247–260. Cardiff: Japan Policy Research Institute.
Sellek, Y. 2003. Migration and the nation-state: Structural explanations for emigration from Okinawa. In *Japan and Okinawa: Structure and subjectivity*, ed. Hook, G.D. and R. Siddle, 74–92. New York: RoutledgeCurzon.
Shimabukuro, J. 1997. Okinawa kenmin tōhyō ni okeru seiji katei [Political process in the Okinawa prefectural plebiscite]. *Toshi Mondai* 89 (2): 23–34.
———. 2015. Okinawa Jiritsu no kōsō to higashiajia no kōzōtenkan [The concept of Okinawan autonomy and the structural transition of East Asia]. In *Okinawa ga tou nihon no anzen hoshō [Revisiting Japan's security from Okinawa]*, ed. Shimabukuro, J. and K. Abe, 287–320. Tokyo: Iwanami Shoten.
Shiohira, Y. 1997. Okinawa kenmin tōhyō, sono go [The Okinawa prefectural referendum and after]. *Toshi Mondai* 89 (2): 67–78.
Smith, S.A. 2003. Beyond the "post-cold war"? U.S.-Japan alliance and the future of Asia-Pacific security. In *Reinventing the alliance: U.S.-Japan security partnership in an era of change*, ed. Ikenberry, J.G. and T. Inoguchi, 117–140. New York: Palgrave Macmillan.
Smits, G. 1999. *Visions of Ryukyu: Identity and ideology in early-modern thought and politics*. Honolulu: University of Hawai'i Press.
Soeya, Y. 2005. Japanese security policy in transition: The rise of international and human security. *Asia-Pacific Review* 12 (1): 103–116.
Taira, Y. 2012. *Sengo okinawa to beigun kichi: "Juyō" to "kyozetsu" no hazama de 1945–1972 [Postwar Okinawa and U.S. military bases, 1945–1972]*. Tokyo: Hōsei Daigaku Shuppankyoku.
Takara, T. 2005. Dairishomei soshō: Heiwa wo uttaeta saiban [The lawsuit of substitute signature: The trial of appealing peace]. In *Okinawa wo heiwagaku suru! [Doing peace studies in Okinawa!]*, ed. Ishihara, M., H. Nakachi and D.C. Lummis, 89–102. Kyoto: Hōritsu Bunkasha.
Tanji, M. 2006. *Myth, protest and struggle in Okinawa*. London: Routledge.
The Committee for the Promotion of Decentralization. 1997. *Dai san-ji kankoku: Bunken gata shakai no sōzō* [The third recommendation: The creation of decentralized society]. Tokyo:

Cabinet Office, Government of Japan. Retrieved from http://warp.ndl.go.jp/info:ndljp/pid/8313852/www8.cao.go.jp/bunken/bunken-iinkai/3ji/.
Toriyama, A. 2013. *Okinawa/Kichi shakai no kigen to sōkoku [Okinawa/the origin and conflict in the base society]*. Tokyo: Keisō Shobō.
———. 2015. Kichi to teikō [Bases and resistance]. In *Okinawa ga tou nihon no anzen hoshō [Revisiting Japan's security from Okinawa]*, ed. Shimabukuro, J. and K. Abe, 111–138. Tokyo: Iwanami Shoten.
Tsuruya, K. 2017 (November 18). *Perī no kokuhaku: Moto kokubōchokan Okinawa eno tabi* [Perry's confession: the former defense secretary's trip to Okinawa] [Television broadcast]. Tokyo: NHK.
U.S. Department of Defense. 1990. *Report to congress a strategic framework for the Asian Pacific rim: Looking toward the 21st century*.
———. 1992. *A strategic framework for the Asian Pacific rim: Report to congress*.
———. 1993. *Report on the bottom-up review*.
———. 1996. Secretary of defense Perry visit to Tokyo, 14–15 April 1996 [Background Book]. In *Digital National Security Archive, DNSA Collection, Japan and the United States: Diplomatic, security, and economic relations, Part III, 1961–2000*. Retrieved from https://search.proquest.com/docview/1679142620?accountid=14891.
U.S. Department of Defense, Office of International Security Affairs. 1995. *United States security strategy for the East Asia-Pacific region*.

The Political Dynamics and Impacts Surrounding Subic Naval Base in the Philippines

Ayae Shimizu

Abstract The Philippines hosted U.S. military bases until their closure in 1992. Subic Naval Base, the largest among them, was regarded as the most important for the security of the Asia-Pacific region. Olongapo, the city adjacent to the base, was heavily impacted by the presence of the base. The people of Olongapo continually protested the negative impacts they suffered due to the existence of the base. This sub-state actor did not directly participate in any negotiation with the U.S. government for its autonomy, but the local voice was strongly manifested in anti-base movements, and they had a pervasive influence on the negotiations between Washington and Manila over U.S. military bases. In this chapter, I explore the issues Olongapo faced, how its people raised a voice of opposition, and how that local voice ultimately facilitated the withdrawal of the U.S. military forces from the Philippines.

Keywords Philippines · Sub-state actor · Subic naval base · U.S. military bases · Autonomy

1 Introduction

The U.S. maintained military bases in the Philippines until 1992. These bases had the longest history of any U.S. bases in Asia, and until their removal, contributed to the security of the Philippines and the Asia Pacific region. During that history, the U.S. and the Philippines conducted many negotiations pertaining to the bases. After World War II, the Military Base Agreement (MBA) enacted in 1947 served as the legal grounds for keeping the American military bases in the Philippines after it had gained independence from the U.S. Covered in more detail hereafter, the MBA was extremely unfair to the Philippines, and it took intense base policy negotiations over many years between the two countries to rectify the agreement to make it more balanced.

A. Shimizu (✉)
The Institute of International Affairs, Tokyo, Japan
e-mail: ayaeshimizu@lion.ocn.ne.jp

M. Takahashi (ed.), *The Influence of Sub-state Actors on National Security*,
Springer Polar Sciences, https://doi.org/10.1007/978-3-030-01677-7_7

Through the MBA, 16 military bases were established in the Philippines.[1] Among them, the largest military facility in terms of area was Subic Naval Base, located in the province of Zambales in the mid-west section of the island of Luzon. Originally, the base completely encompassed Olongapo City. However, the Philippine government was eventually able to get a large portion of the land returned to them from the U.S. government through negotiations. Even so, the base took up 6658 hectares as of 1990.[2] Subic had numerous negative impacts on the people on the base and in the surrounding areas. In addition to Subic, another well-known base named Clark Air Base was located in Angeles City in the province of Pampanga. As with Olongapo City, Angeles City was a sub-state actor which was also heavily impacted by hosting a base. In fact, all the sub-state actors where U.S. military facilities were built, were negatively influenced by the bases. There were many issues that the locals faced in their dealings with the U.S. military, such as incidents and accidents that occurred in and around the bases. Other problems included employment and environmental issues (Calder 2007). This is especially true for the large-scale bases in Olongapo and Angeles, the impacts of which were considerably stronger than those of the smaller U.S. bases on other sub-state actors.

Research into how these sub-state actors in the Philippines influenced the U.S. military is still insufficient. The focus of prior studies dealing with base politics and the U.S. military bases in the Philippines has focused mainly on the negotiations between Washington and Manila. The primary focus of research has been the "special relationship" between the U.S. and the Philippines (De Castro 2003). What most studies deal with is how the U.S. supported the Philippines militarily in its domestic struggles, in exchange for maintaining U.S. bases in the Philippines. In other words, the inequitable terms that the Philippines afforded the U.S. is what most research has been done on. What has not been examined thoroughly is the role that sub-state actors played in compelling the changes in the relationship between the U.S. and the Philippines. A focus on these sub-state actors is needed to understand how they changed the political environment and ultimately lead to the U.S. military pulling out of the Philippines.

This chapter will examine how sub-state actors were impacted by the U.S. military bases in the Philippines. Although there were other military facilities in the Philippines as noted above, I shall focus on Olongapo City for the following two reasons. First, it had the largest base and was the most severely influenced. The U.S. emphasized that Subic Naval Base was strategically the most important base in the Philippines as it was the largest U.S. military facility in the country, and, as will be described later, had an extremely important strategic position for the U.S. as a supply and repair base (Yeo 2011). However, there were many incidents and accidents associated with the base (i.e., the fatal shootings of Filipinos and mistaken firing

[1] Agreement between the United States of America and Republic of the Philippines Concerning Military Bases, 14 March 1947.

[2] Simbulan, R. 2009. *Forging a Nationalist Foreign Policy: Essays on U.S. Military Presence and the Challenges to Philippine foreign policy*, Ibon, Quezon City: 2.

during the training by the U.S. military), which is why anti-base protests frequently occurred.

Second, at the time the Philippines gained its independence from the U.S., "the city of Olongapo itself was a U.S. base." As was stated in the first paragraph of this section, the U.S. and the Philippines entered into the MBA in 1947, and the agreement acknowledged U.S. sovereignty and control over not only the base and the U.S. military facilities, but also Filipino laborers, criminal justice, and all of Subic Bay and Olongapo City (Cooley 2008). Therefore, the U.S. military base dramatically impacted Olongapo City. However, it is also true that the city was able to recover its autonomy later and have a major influence on the U.S. military bases in the Philippines. This chapter will detail the events in Olongapo City, specifically, how protests against the base and anti-base activities affected base politics and influenced the U.S. to withdraw from the base.[3]

The following section will shed light on the process of the establishment of the MBA in 1947, the document that provided the legal grounds for the U.S. military to maintain its bases in the Philippines, as well as on the occurrence of base-related problems during the Cold War period, and the revision of the MBA. The section will also examine how Olongapo City recovered its autonomy. In the subsequent section, the revision of the MBA due to the increased strategic value of Subic Naval base after the Vietnam War and the influence that the anti-base activities in Olongapo City had on the democratization of the Philippines and the enactment of the new constitution will be discussed. The third section clarifies the influence of Olongapo City on the refusal in the Philippine parliament to ratify the treaty and the U.S. withdrawal in the 1990s.

1.1 The Roots of U.S. Military Bases in the Philippines

The history of U.S. military bases in the Philippines goes back to 1898, when the U.S. acquired the islands during the Spanish-American War. After fighting Spain, the U.S. took possession of the Philippines, Guam, and Puerto Rico. At that time, the US was attempting to expand its presence in Asia through imperialism influenced by Alfred Mahan's "Ocean Power Theory." (Mahan 1906) One such imperialist was Theodore Roosevelt, who at the time was Assistant Secretary of the Navy. He felt the need to expand the navy and obtain colonies overseas (Baker 2004). The U.S. established a naval base at Subic in the province of Zambales, making it America's first base in the Asia Pacific region. Subic had already served as a naval base for Spain and was ready for immediate use. However, at the beginning of the twentieth century, Hawaii was the base of choice for the Asia Pacific region, with

---

[3]About protest activities and anti-base movements, see Kawato, Y. 2015. *Protests against U.S. Military Base Policy in Asia: Persuasion and Its Limits*, Stanford University Press, Stanford, California: 14–15.

military functions established there in 1903. Thus, a dockyard for repairs was built at the Subic base, which later came to be seen predominantly as a repair base.

In the following decades, the Philippines was left to the whims of the U.S.-Japan relationship. Japan won the Russo-Japanese War in 1905 and established itself as a military power. To the U.S., Japan had become a hypothetical enemy. The Philippines were in geographical proximity of Japan, and the U.S. military assessed that it would be extremely difficult to protect the Philippines from Japan's strong navy.[4] It was clear that the Philippines would become a battleground in the event of a war between the U.S. and Japan, and both the U.S. and Japanese governments avoided points of conflict that would inflame their relationship. In the U.S., there were increasing thoughts of giving up the Philippines, and even Roosevelt, who had annexed the Philippines, began talking about an early withdrawal.[5]

The Great Depression, which began in 1929, as well as the Manchurian Incident in 1931 accelerated discussions in the U.S. about an early withdrawal from its island colony. In 1934, self-governance began in the Philippines, with full independence to come after, approximately, 10 years of preparation. Thus, the Commonwealth of the Philippines' government began in 1935, and in 1946, the preparations for independence were complete.

Negotiations regarding the U.S. bases in the country began in 1943. Manuel Quezon, the president of the Commonwealth of the Philippines, sent a document on the treatment of the U.S. bases in the Philippines after independence to the U.S. president Franklin D. Roosevelt, to prepare for the event. He noted that it would be necessary to keep U.S. forces in the Philippines to preserve the peace not only for the Philippines, but the entire Asian region.[6] Quezon suddenly died from a disease in 1944, and his Vice President Sergio Osmeña was elevated to President. Osmeña was supportive of the U.S. - in his frequent visits to Washington, D.C. he repeatedly stated the need to keep the U.S. military in the Philippines,[7] and did his utmost to that end.

However, in 1945 Vice President Harry S. Truman become the president of the U.S. after the death of Franklin D. Roosevelt, at which time, opinions regarding U.S. bases in the Philippines diverged in two directions. The first was that, given the inevitable withdrawal from the Philippines, bases should be kept to a minimum size. The second was that U.S. bases in the Philippines should be maintained to keep Asia stable. In May 1945, the first formal negotiations regarding the bases were held

---

[4] Nakano, S. 2003. "*Philippine no beigun kichi mondai: Shokuminchi jidai kara 1992-nen made*" [The Issue of US Bases in the Philiippines: From the Colonial Era to 1992], ed. by Hiroshi Fujimoto & Masashi Shimakawa, "*America no sensou to zainichi beigun: Nihon no anzen hoshou taisen no rekishi*" [Us Wars and the US Military in Japan: A History of Japanese Security], Shakai hyouronsha: 166–167 (In Japanese).

[5] Ibid.

[6] *Foreign Relations of the United States*[hereafter cited as *FRUS*]1945, Vol. VI: 1203–1205.

[7] Berry, W. 1989. *U.S. Bases in the Philippines; The Evolution of the Special Relationship*, Westview Press: 13–14.

between Truman and Osmeña in Washington, D.C.[8] In advance of those talks, the U.S. prepared a plan regarding its military bases - the plan acknowledged the right of the U.S. military to freely come and go with regards to the U.S. naval and air bases in the Philippines. It stipulated exemptions from taxes for supplying goods to U.S. bases, as well as the U.S. right to secure labor for operating and protecting the bases. In this way, the provisional base agreement came to reflect the details stated in the planning document mentioned above, prepared by the U.S.[9] The Philippine leadership's strong desire for the U.S. military to stay compelled them to assent to a provisional agreement that reflected the colonial rights of the U.S.

1.2 The 1947 U.S.–Philippines MBA

The MBA, concluded in 1947 between the U.S. and the Philippines, set forth borders for military bases with land provided by the Philippines to the U.S. It stipulated construction, control, and usage of the bases by the U.S. military. There were 16 U.S. military bases in the Philippines, including Subic Naval Base and Clark Air Base. All rights regarding these bases were held by the U.S. Of particular importance was that the U.S. authority extended not only to the U.S. bases and military facilities, but to Filipino laborers working in those facilities, criminal jurisdiction, and the city of Olongapo itself.[10] In other words, it was acknowledged that the U.S. had sovereignty in both Subic Bay and Olongapo City. The foreword to the MBA emphasized the significance of these rights and responsibilities and especially emphasized that criminal jurisdiction was of the utmost importance.[11]

When considering base politics in the triangle between the U.S., the Philippines, and Olongapo City, the 1947 MBA was critical in terms of the relationship between these three entities. Later, to overcome issues related to the bases, the U.S. and the Philippines repeatedly negotiated for more than 30 years to revise the MBA. These revisions were spurred by the local governments hosting U.S. bases, including Olongapo City. More than any other municipality, Olongapo had its sovereignty and control taken from it entirely. This is the primary reason it continually objected to the MBA, wishing to have that sovereignty restored.[12]

---

[8] *FRUS*, 1945: 1208.

[9] Berry, 1989: 47–68. Also, see the following. Nakano, S. 2007. *Rekishi keiken toshite no beikoku teikoku: Beihikankeishi no gunzou* [*The US Empire as Historical Experience: the Group Images of the History of the US-Philippines Relationship*], Chap. 6, Iwanami Shoten (In Japanese); Ito, Y. 1998. "Philippine no senryakuteki juuyousei no henka to 1947-nen beihi gunji kyoutei no seiritsu katei" [Changes in the Strategic Necessity of the Philippines and the Process of Creating the 1947 US-Philippines Military Agreement], *International Relations*, No. 117, The Japan Association of International Relations (In Japanese) .

[10] Cooley, 2008: 66.

[11] Berry, 1989: 47.

[12] Kawato, 2015: 125–130.

As was noted in the previous paragraph, since the time of its independence, Manila had been asking the U.S. government to maintain its military forces and bases in the Philippines, because it did not have sufficient power to maintain its own security. The MBA was enacted as a way for the U.S. to respond to these requests from the Philippines. Accordingly, regarding the MBA, the Philippines' central government was forced to make many concessions to put a halt to the U.S. military withdrawals. That made the MBA an extremely one-sided document. In addition to the U.S. being able to use the military bases practically free of conditions and with no distinction between emergencies and peacetime, criminal jurisdiction therein was also given to the U.S. Because of this, the Philippines could not prosecute any crimes committed by U.S. soldiers against Filipinos. In addition to such inequities, the MBA was disadvantageous to the Philippines in other ways – for example, the time limit for the usage of the bases was set at 99 years, which was tantamount to there being no limit.[13] The MBA had negative impacts on Olongapo and ultimately changed its history.

1.3 Emergence of Base Problems and the Return of Policing Rights to Olongapo

Not long after the MBA was entered in 1947, problems began to surface in areas surrounding U.S. military bases, including Olongapo City. Public anxiety and dissatisfaction with the U.S. military bases grew, though these emotions were mostly directed toward the unfair MBA, rather than the bases themselves. The Filipino public understood that a great gap existed between the MBA and the Status of Forces Agreements concluded between the U.S. and other countries, and thus they focused their anger on the unfair provisions of the MBA, while still recognizing the importance of the bases.

The issue the citizens of Olongapo City were particularly dissatisfied with were the arrangements regarding criminal jurisdiction for crimes caused by U.S. soldiers. For example, around 1953, the NATO Status of Forces Agreement accorded the U.S. the right to prosecute U.S. soldiers for crimes committed on NATO soil only in the event the crimes threatened the property of the U.S., the victims were U.S. citizens, or the crimes were committed while the soldier was on duty. Outside of these cases, the countries where the U.S. bases were located had the right to prosecute crimes committed by U.S. soldiers.[14] Having criminal jurisdiction meant the countries were recognized as independent states.[15] U.S. soldiers frequently committed

[13] See the following, published by the Foreign Service Institute of the Philippines, for the 1947 MBA in its entirety. Castro, P. A. 1985. ed., *Agreements on United States Military Facilities in Philippine Military Bases 1947–1985*, Foreign Service Institute, Manila, Philippines.

[14] Berry, 1989: 57–60. Nakano, 2003: 166–167 (In Japanese).

[15] Berry, 1989: 62.

crimes in Olongapo City, and the soldiers' ability to escape punishment for their crimes almost incentivized incidents, including theft, assault, rape, and murder.

U.S. Vice President Richard Nixon attended the 10th anniversary of the Philippines' independence in July 1956. An amendment to the MBA had been proposed at the U.S.-Philippines talks held in 1954, and both sides agreed to negotiate toward that end. When those negotiations began, their focus was on the U.S.' obligation to provide security under the U.S.–Philippines mutual defense treaty, an agreement from prior consultations, and the reduction of the MBA expiration period from 99 years to 25 years.[16] However, the negotiations soon fell apart, mostly because the then U.S. president Dwight Eisenhower took a passive stance regarding the U.S. bases in the Philippines, as he held a negative view of U.S. bases in that country. It was the aftermath of World War II, so the threats to the security of the Asia- Pacific region were somewhat mitigated. Consequently, Eisenhower was reluctant to accept the demands of the Philippines government, which asserted equal standing with the U.S. in an amendment to the MBA (Sandars 2000).

Due to pressure from the Philippine Senate in 1958, the Philippines government asked the U.S. administration to reopen negotiations to amend the MBA. The U.S. showed a certain level of empathy and made concessions during those negotiations. The negotiations between U.S. ambassador Charles Bohlen and the Philippines Minister of Foreign Affairs, Representative Felixberto Serrano culminated in the adoption of the amendment in October 1959. This amendment came to be known as the Bohlen–Serrano Agreement.

The following are the main points of agreement in the amendment.[17] (1) Establishment of a prior consultation system in the case the bases needed to be used for purposes outside of the U.S.-Philippines Mutual Defense Treaty, or SEATO, and for bringing long-range missiles into the bases. (2) The reduction of the MBA expiration term from 99 years to 25 years as a condition for bringing the items agreed upon between the two countries into effect. The term could be shortened further upon the agreement of both governments. (3) Olongapo, which had been under U.S. Navy control since 1920, except for the war years, was returned to Philippine jurisdiction. Of special note was the return of exclusive jurisdiction over Olongapo City through the amendment. Twelve years following the signing of the MBA, Olongapo City at long last had its right to police itself restored. Additionally, the flag of the Philippines now flew alongside the U.S. flag at Subic Naval Base, and in all other U.S. military facilities.[18] The days of being completely dominated by the United States were now over and the position of the Philippine government in relation to the bases had become clearer.

[16] Nakano, 2003: 182.

[17] Military Bases in the Philippines, Relinquishment of Olongapo and Adjacent Area, Agreement Between the United States of America and Republic of the Philippines, 7 December 1959, *Treaties and Other International Acts Series (TIAS)*, Washington, DC: Department of State: 4388.

[18] See the following publication by the Philippines' Department of Foreign Affairs, for the full text of the Bohlen-Serrano Agreement. Castro, P. A. 1985. ed., *Agreements on United States Military Facilities in Philippine Military Bases 1947–1985*, Foreign Service Institute, Manila, Philippines.

Despite the amendment to the MBA, the anti-base movement in Olongapo became increasingly strident in the 1960s. In addition to the fact that the crimes perpetrated by U.S. soldiers in Olongapo City did not diminish, the issue of criminal jurisdiction was not resolved. The U.S. attempted to suppress the growing anti-base sentiment among the citizens of the Philippines by returning base land, but there was still no resolution of the question of criminal jurisdiction, which was a major factor in the opposition to the bases. Because of this, every time a U.S. soldier committed a crime the inequity of the MBA came to the forefront, and the emotions of the populace were stoked. In 1964 a Filipino that had approached a Subic munitions depot in a small boat was shot as an aggressor.[19] As similar shootings of Filipinos by U.S. soldiers came to light, the anti-base sentiment of the populace reached its peak. A large demonstration involving more than 2000 participants took place in Angeles City, where Clark Air Base was situated, and where similar shootings had occurred.[20]

The government of the Philippines asked the U.S. government to amend the agreement to resolve the situations in Olongapo City and Angeles City. Washington responded to that request, and in August 1965 the U.S. ambassador, William McCormick Blair, and Philippines' Secretary of Foreign Affairs, Mauro Mendez, signed an agreement that, in part, amended the MBA. This became known as the Blair–Mendez Agreement. The agreement acknowledged that criminal jurisdiction rested with the Philippines for crimes committed by U.S. soldiers who caused damage to Filipinos, including crimes committed by off-duty soldiers. The provision concerning criminal jurisdiction was formally added to the MBA through the Rusk–Ramos Agreement. This agreement at last recognized criminal jurisdiction and brought the MBA up to the level of the NATO and the Status of Forces Agreement between the U.S. and Japan.[21]

1.4 The Post–Vietnam War Importance of Subic Naval Base

U.S. military bases in the Philippines housed 27,000 soldiers during the Vietnam War. Afterwards, that number dropped to 15,000 by 1976. How the U.S. military would continue to station soldiers in the Philippines came into question. However, the U.S. military bases in the Philippines were of growing strategic importance, owing to Pacific and Indian Ocean routes and the Soviet Union that used them. While declaring that "the primary duty of U.S. bases in the Philippines is to protect the maritime routes of free nations from the threat of the Soviet navy," Joint Chief of Staff Chairman, George S. Brown, assessed that the bases enabled mobility of the U.S. Seventh Fleet by uniting it with the Diego Garcia atoll in the Indian Ocean.[22]

[19] Asano: 183.

[20] Kawato, 2015: 128–129.

[21] Berry, 1989: 104–107; Asano: 183.

[22] *Asahi Shimbun*, February 21, 1978.

At the same time, Washington believed that if U.S. bases in the Philippines became the property of the host country, then their functioning would be seriously constrained and a decrease in the strategic importance of the bases would be unavoidable, causing damage to the U.S.' wide-area strategy. At the middle of the 1970s, when the U.S. military had been deployed in Thailand, the Thai military did not allow a free use of the bases by the U.S. military and demanded priority usage rights in areas where the U.S. military was located. Thus, the U.S. military realized that the bases in the Thai territory were inefficient, which lead to a withdrawal from Thailand. Therefore, the inherent strategic value of bases in the Philippines grew, and this made it even more difficult for the U.S. to cede base rights to the Philippines.

In particular, Subic Naval Base held tremendously strategically importance. Comprising 10,500 hectares of ocean and 15,500 hectares of land, the base was positioned as the largest repair and resupply base in the U.S.' Seventh Fleet, for both the western Pacific and Indian oceans. As of 1977, Subic Naval Base conducted 60% of the repairs for the Seventh Fleet and was always ready to handle approximately 15% of the fleet for repairs. A day of labor for repair work was worth $88 at Yokosuka, $142 in Guam, and $158 on the West Coast of the U.S., but only $21.50 at Subic.[23] Therefore, Subic Naval base was the optimal facility for the U.S. geopolitically, strategically, and financially.

Thus, the U.S. bases in the Philippines came to be highly valued, not only as supply and training bases, but also strategically, because they were front-line bases in the western Pacific Ocean amid the increasing activity of the Soviet navy in the Indian Ocean. Additionally, after the conclusion of the Vietnam War, the aggressive involvement of the Soviet Union in Vietnam was of great concern to the U.S. military. The former U.S. military base at Cam Ranh Bay became the Soviet Union's largest overseas base after the Vietnam War, and the U.S. were determined not allow the bases in the Philippines to fall out of their hands.

Although some portions of U.S. bases in the Philippines were returned by 1977, Subic Naval Base remained the largest naval base west of Hawaii.[24] U.S. military bases in the Philippines defended and secured sea lanes in the western Pacific together with U.S. bases in Japan. Thus, they remained critical, especially since the distance from Manila to Guam is only 2800 kilometers, and from Manila to the Diego Garcia atoll is only 6500 kilometers, which makes the Philippines a very important relay point between the Pacific and Indian Oceans.

[23] Asano, Y. 1978. "Beihi kankei no kongo wo saguru" [Exploring the Future of US–Philippines Relations], Asia Trends No. 4: 64–78 (In Japanese).

[24] Ibid.: 64–78.

1.5 The 1979 Base Agreement Amendment and U.S. – Philippines Relations in 1980s

The end of the Vietnam War in 1975 was one of the elements justifying a U.S. withdrawal from the Philippines. Nonetheless, President Gerald Ford agreed to further negotiations regarding the U.S. bases during his visit to Manila in December in 1975. A new round of negotiations started in 1976. The U.S. sought to demonstrate their resolve to remain in Asia following the withdrawal from Vietnam and to show a clear recognition of Philippine sovereignty. President Ferdinand Marcos used the situation to try to maximize the economic assistance the U.S. would provide to the Philippines.

There was also concern that, if the U.S. did withdraw, a power vacuum would be created, allowing some other country to extend its influence (In fact, China did ask Manila to use bases in the Philippines after the U.S. withdrawal). It was in such circumstances that President Marcos made a visit to China in 1975 and established diplomatic relations between the two countries. Furthermore, Marcos leveraged the prospect of drawing closer to the USSR in base agreement amendment negotiations with the U.S. to get more economic aid. The response of the Ford administration, which had shown a willingness to draw closer to the Philippines, despite being confronted with the Philippines' unpredictability, was to try to reduce the tension between the two countries. By acknowledging the strategic value of its military bases in the Philippines, the U.S. found itself compelled to keep its bases there. Late in 1976, Henry Kissinger made an offer of one billion dollars in aid over 5 years for the retention of the bases. However, Marcos refused the offer. The negotiations then paused because of the U.S. presidential election. The new Carter administration resumed negotiations with Philippines. Richard Holbrooke, Assistant Secretary of State for Asia, was tasked with handling the negotiations.

As a result, the MBA was amended in 1979. The amended agreement was signed by Carlos Romulo, the Foreign Minister of the Philippines, and Richard Muphy, the U.S. Ambassador to Manila, on 7 January 1979. Reductions of the base at Subic Bay and Clark Airfield were agreed to. Particularly, Clark Airfield saw 90 percent of the overall base area returned to the Philippine authorities. All bases occupied by U.S. forces were recognized for the first time as military facilities of the Philippine government. The Philippine flag was to be flown at the bases and Filipino commanders were to be appointed with responsibility for perimeter security at the bases. Although Philippine sovereignty was recognized in all bases in the country, U.S. commanders continued to direct the U.S. military forces. The U.S. government considered the Philippine government responsible for maintaining the bases in the Philippines.

In this amendment, the MBA would be reviewed every 5 years, the Philippine flag would fly on bases, Philippine commanders (with limited authority) would be stationed at each base, perimeter security would be provided by Philippine forces. The Americans promised to hold suspects in the Philippines for a reasonable amount of time. Furthermore, the U.S. provided half a billion dollars as "best effort" in aid to the Philippines over 5 years. The door closed on opposition to continuing the bases after the end of the Vietnam War and the U.S. aimed to restructure the order in the Asia region by maintaining its bases in the Philippines.

Marcos won the presidential election for the third time in an overwhelming victory in 1981. U.S. Vice President George Bush attended Marcos' inauguration ceremony, and the following year, Marcos visited the U.S. as a guest of the state. As the Reagan administration and Marcos administration built close relations, in 1983, the MBA was again amended as part of the "five-year review" set forth in the 1979 MBA amendment. Philippine Ambassador Benjamin Romualdez and U.S. Ambassador Michael Armacost reached an accord in June 1983 in Washington. They reconfirmed that sovereignty of the bases rested with the Philippines.[25]

Moreover, the U.S. agreed to make the "best effort" payments of 900 million dollars over 5 years that Marcos had requested (Paez 1985). The amendment reconfirmed the importance of the bases in the Philippines. The U.S. government again acknowledged that Subic Naval Base was a strategically important base for security in the Asia-Pacific region.[26] The U.S. agreed to notify the Philippine government of any major changes in force levels, equipment, or weapon systems. The U.S. pledged continued efforts to improve the conditions in areas surrounding the bases, particularly in Olongapo City and Angeles City.[27]

1.6 Philippines Democracy and the Enactment of a New Constitution

Soon after the agreement was signed, political events started to unravel in the Philippines. Benigno Aquino, the opposition figurehead and the political rival of Marcos, was assassinated in August 1983. After his death, the democracy movement started to gain momentum in the Philippines, and the dictatorship of Marcos began to crumble. In 1984 Marcos went into seclusion to recover from illness, while political dissidents, opposition forces, and anti-base activists started to rally around Aquino's widow, Corazón Aquino, as the new opposition figurehead.

Naturally, the U.S.–Philippines relationship worsened. Filipinos castigated the U.S., which had supported the Marcos dictatorship in order to keep the U.S. bases. Dissatisfaction with the unfair MBA among Filipinos was directed at the U.S. military bases of Subic and Clark. In February 1986, the Marcos administration was brought down by the "People Power Revolution," which ushered the Corazón Aquino administration into office.

---

[25] Defense Memorandum of Agreement, *TIAS* 10699.

[26] Central Intelligence Agency; CIA, "The Key Role of U.S. Bases in the Philippines," reported by Asian Study Center, January 10, 1984, https://www.cia.gov/library/readingroom/docs/CIA-RDP88T00528R000100010046-1.pdf. (accessed April 20,2018).

[27] Sandars, 2000: 115.

President Aquino had been swept into power by a popular revolution hostile to the Marcos regime and democracy took hold of the country, but the political situation in the Philippines continued to be unstable. Demonstrations by authoritative opposition organizations could not be handled by the Philippines' military, and the government relied on the U.S. military to ensure security for the country.

In May 1986, President Aquino created a Constitutional Commission that would draft a new constitution. The Constitutional Commission completed its work in October 1986. Aquino submitted the constitution to a plebiscite in February 1987 where it was approved by the majority of voters. The constitution included some provisions that would have significant implications for the base policy in the country. First, in anticipation of the expiration of the MBA in 1991, the new constitution granted the Senate the final authority in deciding the fate of military bases.[28] The government of the Philippines adopted the new constitution by popular vote, banning the bringing of nuclear weapons into and stationing of foreign troops in the country. Article 18, paragraph 25 of the new constitution stated the following:

> After the expiration in 1991 of the Agreement between the Republic of the Philippines and the United States of America concerning military bases, foreign military bases, troops, or facilities shall not be allowed in the Philippines except under a treaty duly concurred in by the Senate and, when the Senate so requires, ratified by a majority of the votes cast by the people in a national referendum held for that purpose, and recognized as a treaty by the other contracting State.[29]

Due to the provisions in this paragraph, a new MBA could no longer be created without the concurrence of the Senate of the Philippines.

The nuclear weapons issue also had become more contentious from the U.S. perspective.[30] In the 1980s some senators in the Philippines held nationalist beliefs and opposed constructing a nuclear plant in the Philippines. They protested the stationing of nuclear weapons in the military bases, too. The U.S. maintained the policy of neither confirming nor denying the presence of nuclear weapons in its military bases, but the prevailing view was that the U.S. military kept nuclear weapons in the Philippines despite the Philippine constitution and legislation that declared freedom from nuclear weapons.[31]

Many new anti-base groups developed during the 1980s.[32] These groups gained momentum due to the Philippine democratization in 1986, the constitutional revisions in 1987, and the renegotiation of U.S. bases under the Philippine American Cooperation Talks (PACT) in 1990–1991.[33] The anti-base groups formed after 1987 connected elites opposed to U.S. bases. That is to say, anti-base activists operated together with the anti-base government elites to try to establish a new era of

[28] Kawato, 2015: 168.

[29] The 1987 Constitution of the Republic of the Philippines, Article XVIII, Sect. 25.

[30] Berry: 291–293.

[31] Berry: 291–293.

[32] Yeo, 2011: 42–49.

[33] Ibid.: 42–49.

Philippine security free from U.S. bases.[34] The new constitution, which stipulated that foreign troops could not be accepted without the Senate's approval after the MBA expired in 1991, was enacted. Thus, an opportunity arose for Olongapo to have an impact on whether the base in the Subic Bay would be retained or not.

1.7 Factors for Withdrawal: Expiration of the MBA and a New Treaty

Since 1987, the Philippine Senate had created legislation to strengthen the constitutional declaration of freedom from nuclear weapons. That legislation gave the anti-base senators added legal basis for defending their position in 1991.[35] The direct factor behind the withdrawal of U.S. military bases in the Philippines from 1991 to 1992 was the expiration of the MBA.[36] The MBA of 1947 defined an effective term of 99 years, but in the Bohlen–Serrano Agreement, signed in 1959, it was agreed that this term should be shortened to 25 years.[37] Thus, the expiration of the MBA was set to 1991 in the Rusk–Ramos Agreement of 1966. As already mentioned, the new constitution, enacted in 1987, stated that, after the MBA expired in 1991, no further modifications could be made without the concurrence of the Senate.

As the expiration of the MBA approached, the U.S. began to take action regarding its bases in the Philippines. A special ambassador for base problems was appointed in the U.S. government: Richard Armitage. In May 1990, the first discussions on base issues took place in a meeting between Philippines' Secretary of Foreign Affairs, Raul Mangrapus, and Armitage, where Mangrapus notified the U.S. of the Philippines' government's position that the validity of the MBA could not be extended beyond September 16, 1991. Conversely, the U.S. government asserted that notification could be made after the MBA expired on September 16, 1991, and that a grace period should be granted for the withdrawal of the bases. The focus of these negotiations was on the amount to be paid as base fees (i.e., subsidies) for "best efforts" during the new MBA period. The Philippines insisted that the MBA did not allow for an extension of the expiration date, but rather included the time needed for withdrawal, and that, therefore, a grace period for the withdrawal should be kept as short as possible. Conversely, the U.S. noted that the bases existed at the request of the Philippines government, and asked for a new, renewable base agreement lasting for at least 10 years, because a withdrawal, should the Philippines not desire the continued presence of the bases, would require time.

After these negotiations, the two countries signed a "Treaty of Friendship, Cooperation, and Security" on August 27, 1991, right before the expiration of the

[34] Ibid.: 43.

[35] Ibid.

[36] See Nakano for the timeline of amendments to the MBA and changes to base expiration.

[37] For more information on MBA negotiations, see Capulong, V. S. 1987. *Philippine foreign policy toward the U.S. 1972–1980: Reorientation?* University of Stockholm: 113–116.

MBA. The term for the new base agreement could be renewed for another 10 years, just as the U.S. asked, and the "best effort" letter promised 2.2 billion dollars in aid over the period of 10 years. Thus, it may be said that, in the end, the Philippines signed a draft treaty practically giving concessions to the U.S.

However, in Olongapo City, anti-base protests had intensified. The protestors wanted the removal of U.S. bases. What is more, they wanted political parties to add issues to their agenda such as democratic reforms and the end of the U.S. security umbrella. The fact that the protests were related to the national security agenda is very important. However, the societal importance of the naval base in Olongapo City was apparent. As Andrew Yeo pointed out, the security consensus among political elites was very weak.[38] Therefore, anti-base political elites tied the base issue together with several other political and economic issues rather than focusing exclusively on the bases.

1.8 Conflict Between the Philippines Government and the Philippine Senate

Secretary of Foreign Affairs Mangrapus and Senate President Salonga held discussions in August 1990, prior to earnest negotiations with the U.S. Although Secretary Mangrapus was opposed to the status quo of maintaining U.S. military bases in the Philippines, he advocated the need to seriously consider issues regarding unemployment and security should the U.S. military withdraw (Bengzon 1997). In the 40 years since the MBA was signed in 1947, the government of the Philippines had continually negotiated with the U.S. about the bases, asking for corrections to items in the MBA that were unfair to the Philippines, attempting to get more favorable conditions. Accordingly, the Philippines' government for a long time did not ask for the withdrawal of the U.S. military so much as it asked for many other things, such as the return of base land, legal jurisdiction for prosecuting crimes and the restoration of sovereignty over the bases. Mangrapus was friendly to the U.S. and felt that good relations could be built by ending their base-centered relationship.

In the meantime, the Senate of the Philippines wanted to end the base relationship with the U.S., as well as the relationship that had been carried over from the colonial period. Senators who supported the withdrawal of the U.S. military felt that the withdrawal itself was a means of satisfying domestic nationalism in the Philippines, and that it would be right and proper.[39] The principle at work tied

[38]Yeo, 2011: 44–45.

[39]See the document entitled "Resolution Expressing the Decision of the Senate not to Concur with the Treaty of Friendship, Cooperation, and Security and Its supplementary Agreements between the Government of the Philippines and Government of the United States, signed in Manila, on August 27, 1991." This and other documents were edited and published in Castro-Guevara, M. (ed.). *The Bases Talks Reader: Key Documents of 1990–91 Philippine-American Cooperation Talks*, Anvil Publishing, Manila: 466–469.

legitimate political power to the aim of a Philippines independent of the U.S. and with no military bases. The new constitution enacted in 1987 clearly stipulated that the MBA would no longer be in effect.[40]

Normative arguments against military bases fit with the knowledge and beliefs of senators who had a more nationalistic view of Philippine history.[41] Their argument was that U.S. military bases were instruments of U.S. neocolonialism and should be closed for the purpose of achieving true Philippine sovereignty and independence. The senators also felt that U.S. military bases must close because the U.S. supported the Marcos's dictatorship. Even after Marcos' fall, these senators believed that the U.S. interest in securing military bases was a form of interference in Philippine politics.[42]

Salonga worked in favor of rejecting the ratification of the treaty, and the Aquino administration that pushed for the ratification of the Treaty of Friendship, Cooperation, and Security came into full conflict with the Senate.[43] The Aquino administration then held a convention, putting pressure on senators to stop their opposition to the treaty ratification, because a withdrawal would put the lives of Philippines' citizens at risk. However, the Senate passed a resolution to "reject the ratification of the Treaty for Friendship, Cooperation, and Security with the U.S." on September 11, 1991, in a 12–11 vote. The 1947 MBA was thus no longer in effect, and in December of that year, the government of the Philippines formally conveyed the end of the MBA to the U.S. government. The following year, the U.S. military fully withdrew from the Philippines.

1.9 Olongapo Opposition to Bases and Base Withdrawal

The intense nationalism caused by the anti-base groups, protests and the climate of democratization intensified after the 1988 MBA review. Around Subic Naval Base, many of the anti-base groups and organizations expressed their opposition to the U.S. military bases and mounted anti-base movements.

According to a CIA analysis, there was no unity regarding the national security agenda in the Philippines, as many people still felt the bases were necessary for national security despite the issues surrounding them. According to the CIA, a lot of Filipinos felt that the U.S. military bases were necessary for the Philippines and that

---

[40] Article XXVIII prohibited the United States from extending the Military Bases Agreement beyond 1991 without the explicit "concurrent" (ratification) by the Philippine Senate and, if requested by the Philippine Congress, an additional popular referendum. Therefore, it was clear that the MBA would not be extended after it expired in 1991.

[41] Kawato, 2015: 138–141.

[42] Ibid.: 139.

[43] See the memoir of Senator Salonga regarding the Senate of the Philippines' attitude toward MBA negotiations: Salonga, J. R. 1995. *The Senate That Said No: A Four-Year Record of the First Post-EDSA Senate Quezon City*, University of the Philippines Press.

they should be maintained. Similar results were found in public opinion surveys, not only in Manila but also in the areas around the bases, including Olongapo City. The CIA also pointed out that anti-base groups tried to involve people who originally did not have a strong opinion on the U.S. bases. The Philippine government, which wanted to maintain the U.S. bases, also had concerns that such people could be persuaded to join the anti-base movement.[44]

Andrew Yeo described how these anti-base groups became allied with the political elites who also opposed the U.S. military bases because they felt the bases weakened Philippine national security.[45] The Philippines had historically entrusted their national security to the U.S. through the Mutual Defense Treaty (MDT), signed between the Philippines and the U.S. in 1951. The MDT is the legal basis for supporting the Philippines in relation to foreign threats. However, the Philippine government also expected internal security issues to fall under the jurisdiction of the MDT. At that time the domestic situation in the Philippines was being destabilized by Islamic anti-government guerrillas (i.e., MILF, Abu Sayyaf, etc.). Despite that, the U.S. thought that internal security in the Philippines was beyond the scope of the MDT. In the eyes of the U.S., the U.S. security umbrella covered the Philippine security with regards to external threats only.

The Philippines faced continued domestic instability, so the anti-base political elites highlighted the American interpretation of the MDT as not applying to the internal security of the Philippines to garner greater support for its anti-base opposition. That realization brought national security anxieties to the forefront. The Philippines never felt directly threatened by the Soviet Union or the communist sphere and the U.S. failed to present a grand strategy that adequately addressed the internal issues that the Filipino people were truly concerned about. Consequently, the absence of perceived external threats and the lack of any clear grand strategy, which also lead to a weakened U.S. political support for the bases, gave the anti-base opposition campaign momentum.[46]

However, not all people in Olongapo City had sought the removal of U.S. military bases. Richard Gordon, who was the mayor of Olongapo City at that time, strongly opposed the withdrawal of U.S. military bases.[47] He particularly feared that the rampant unemployment that would follow the closure of Subic Naval Base would be devastating for the people of Olongapo City. He mobilized Filipino workers that had been working at the Subic base to protest in Manila and was involved in demonstrations seeking to keep the bases active in the Philippines.

Thus, there was no clear consensus regarding the U.S. base in Olongapo. Fierce base-opposition campaigns were taking place, while others like Gordon, the Mayor of Olongapo City, fought to keep the U.S. military bases. As a result of this weak

---

[44] CIA 1987. "The Philippines: Exploring View on the US Military Bases Information as of 24 August 1987" https://www.cia.gov/library/readingroom/docs/CIA-RDP90T00114R000200970001-0.pdf. (Accessed on April 20, 2018).

[45] Yeo, 2011: 44–49.

[46] Ibid.

[47] Simbulan, 2009: 2.

resolve regarding the issue, the nationalist political elite was able to move forward with the legal preparations for the withdrawal of U.S. forces and the return of the bases and surrounding lands to the Philippines.

In the Philippines, the executive and the foreign affairs department were largely in favor of retaining the U.S. bases, especially Subic Naval Base. The U.S. government too recognized that Subic Base was important strategically. In this situation, in the absence of strong resolve, the Filipino political elite with anti-base sentiments responded to activist efforts by embracing the anti-base sentiment cause. The lack of security consensus among anti-base elites in the Congress and Senate posed obstacles for the president and other pro-base groups in forging any cohesive policy to counter anti-base rhetoric.[48] In this instance, a relationship between activists in Olongapo, government elites, and other anti-base movements was formed and collectively was able to push the policy direction of the government toward the withdrawal of Subic Naval Base.

Thus, Olongapo's anti-base movement resulted in the closure of Subic Naval Base. Olongapo, which had been negatively impacted by the base, in turn, influenced the withdrawal of Subic Naval Base with its anti-base movements. In the end, the U.S. government found that President Aquino was able to manage without the U.S. bases. Concurrently, the idea that bases in the Philippines were indispensable was disproven. Thanks to the end of the Cold War and the collapse of the Soviet Union, the functions of Subic Naval Base and the tasks of the naval forces stationed at the Subic Bay were transferred mainly to Japan and Hawaii.[49]

2 Conclusion

This paper has demonstrated that Olongapo City, a Filipino sub-state actor, played an influential role in the withdrawal of U.S. military bases from the Philippines. First, historically, Subic Naval Base had a predominantly negative influence on Olongapo City. At the time the MBA was concluded, Olongapo endured many issues with regards to the base. Filipinos were killed by U.S. soldiers in Olongapo and such incidents occurred repeatedly. Since the 1950s, anti-base movements protesting these incidents had continually gained steam. Filipino people participated in anti-base movements seeking revisions to the unfair and unequal MBA. The Philippine government negotiated with the U.S. government to revise the MBA as a response to these anti-base movements. Through the negotiations, the right to self-police Olongapo City was recovered in the 1950s. After that, the locals protested the MBA in an attempt to have the criminal jurisdiction policies revised. In the 1980s, finally criminal jurisdiction was granted so that the Philippine side could prosecute crimes committed by U.S. soldiers.

[48] Yeo: 62.

[49] Sandars, 2000: 125.

Second, Olongapo City had its autonomy restored, and influenced the negotiations between the U.S. and the Philippine governments. In the late 1980s, parallel with the wave of democratization, the voice of the people that the U.S. military bases represented colonial rule became increasingly felt, and anti-base movements in Olongapo escalated. After the democratization and the enactment of the new Constitution in the Philippines, a political climate that enabled the MBA to expire, was created, with the Philippines deciding not to accept new foreign bases unless congressional approval was given. In addition, Olongapo objected to the fact that nuclear weapons were being brought to Subic, bringing about a law banning nuclear weapons in the Philippines. Supported by the democratization of the Philippines, anti-base movements in Olongapo gained strength and established links with political elites in Manila who opposed the bases. The Philippines Senate refused to ratify the new agreement which was intended to keep the U.S. military and the U.S. military bases in the country after the MBA expired. The ties between anti-base groups and political elites prevailed because the political consensus in the Philippines regarding the need to maintain the presence of the U.S. bases was weak.

Olongapo never directly negotiated with the U.S. government about the naval base in the Subic Bay. However, through its negotiations over the MBA with the Philippine government, the U.S. government became strongly aware of the anti-base sentiment in Olongapo. The U.S. government was forced to make concessions to placate the campaigns opposing the Subic base. The Philippine government, on the other hand, wanting to obtain military and economic assistance from the United States and keep a U.S. presence tended to be dependent on the U.S. government and to lean towards Washington's thinking in intergovernmental negotiations. However, with a more united anti-base movement the Philippine government was forced to change its policies.

Olongapo City suffered numerous incidents, accidents, and crimes as a result of the presence of Subic Naval Base. With regard to these problems, small sub-state actors such as Olongapo City face immense difficulty negotiating directly with major state powers like the Unites States. However, the anti-base movements in Olongapo certainly influenced negotiations between the U.S. and the Philippines. The people of Olongapo city struggled but prevailed in gaining their autonomy – their efforts to regain autonomy were realized gradually over many years. The revisions of the MBA are a result of such sustained work by the anti-base movements and the process of democratization in the country as a whole. The main theme of the present book is autonomy and national security from the viewpoint of subnational entities, and Olongapo demonstrates the influence that a sub-state actor can have on national security. In recent years, Olongapo City has become a special economic zone and has further expanded its autonomy.

References

Baker, A. P. 2004. *American soldiers overseas: The global military presence*, 111–113. Westport: Praeger.

Bengzon, A. R. A. 1997. *A matter of honor: The story of the 1990–91 RP-US bases talks*. Manila: Anvil Publishing.

Calder, K. E. 2007. *Embattled garrisons: Comparative base politics and American globalism*, 83–90. Princeton: Princeton University Press.

Cooley, A. 2008. *Base politics: Democratic change and the U.S. military overseas*, 66. Ithaca: Cornell University Press.

De Castro, R. C. 2003. The revitalized Philippine-US security relations: A ghost from the cold war or an Alliance for the 21st century? *Asian Survey* 43 (6).

Mahan, A. T. 1906. *The influence of sea power upon history, 1660–1783*. Boston: Little Brown.

Paez, P. A. 1985. *Bases factor: Realpolitik of RP-US relations*, 452. Manila: Center for Strategic and International Studies of the Philippines.

Sandars, C. 2000. *America's overseas garrisons: The leasehold empire*, 110. Oxford: Oxford University Press.

Yeo, A. 2011. *Activists, alliances, and anti-U.S. base protests*, 39–40. Cambridge: Cambridge University Press.

Conclusion: The Political Choices of Sub-state Actors and the Politics Surrounding U.S. Military Bases

Minori Takahashi

Abstract The goal of this chapter is to integrate and sum up the discussions from previous chapters and thus offer new insights regarding the role of sub-state actors in the field of security by focusing on the following three correlations:

1. We theoretically extrapolate how the policy choices of subnational actors affect the interaction between the US politics regarding its overseas military bases and the polities of the countries hosting them (The correlation between vulnerability and hold-up);
2. We shed light on how the local voice manifests itself and how it functions in various political matters concerning US overseas military bases at the level of national security by examining the cases of Greenland, Okinawa and Olongapo (The correlation between autonomy and military bases), and
3. We evaluate the relationship between the emergence of Cold-War-like power relations in the post-Cold War period, which are visible in the current Arctic security environment, and sub-state actors (The methodological correlation between deduction and induction).

1 "Paradiplomacy" and the "vertical relationship with the state" as Departure Points

The influence exercised by a sub-state actor on national security, or its attempts to get involved in the field of national security, and by extension, in international relations as a whole, have in the past been termed "second-track diplomacy", while the expression "paradiplomacy" has been especially favored in the field of Arctic studies (Heininen 2014, Kristensen et al. 2018). If we were to take this term literally, it would probably mean "*parallel* diplomacy", but since "para" also connotes meanings such as "irregular", "supplementary" or "quasi", its interpretation may differ significantly from speaker to speaker. Nonetheless, "paradiplomacy" is a term that,

M. Takahashi (✉)
Slavic-Eurasian Research Center and Arctic Research Center, Hokkaido University, Hokkaido, Japan
e-mail: minoritakahashi@slav.hokudai.ac.jp

M. Takahashi (ed.), *The Influence of Sub-state Actors on National Security*, Springer Polar Sciences, https://doi.org/10.1007/978-3-030-01677-7_8

in most usages, denotes international activities of sub-state actors (such as local or regional polities), and is, overall, widely used in a positive sense. In it, various international activities of sub-state actors that are not visible from the viewpoint of "diplomacy", which has strongly tended to be regarded as limited to mutual relations between states, are understood as something conducted "parallel" with the conventional diplomacy.

One such example are the activities of sub-state actors in the Arctic Council (AC), a high-level inter-government body established in 1996. In the AC the eight member states are not the only players, as six organizations of indigenous people are also enrolled as permanent participants and can be seen conducting talks centering on issues that fall in the scope of soft paradiplomacy or low politics with the member states in an atmosphere of partnership based on compromise and mutual agreement. The six indigenous peoples' organizations do not possess the right to make final decisions but influence the political functioning, give legitimacy to the existence of the AC, and are spoken of as important subjects, which is indicative of the existence of paradiplomacy in the Arctic region. On the other hand, such actions by sub-state actors are often interpreted as a consequence of the political stance by the nation states, which seek to lower the political cost and aim for a soft landing in matters that concern them by creating an atmosphere of mutual cooperation. There is even a tendency to view such activities of subnational entities as, ultimately, a dependence on the discretion of the state. How free are sub-state actors in controlling their actions, and can they concretely express their intentions through their actions? That would be very difficult to judge based on the arguments in the past scholarship.

Indeed, the original motive for compiling this volume was a certain discomfort with the frame of reference employed in debates on sub-state actors thus far, that has not managed to break free from the concept of *vertical* relationship with the state, despite the fact that activities deemed as *para*diplomatic, have been vigorously discussed. Furthermore, the scope of paradiplomacy thus far has often tended to be limited to the layer of soft paradiplomacy or low politics, while not enough argument based on sufficient theoretical and empirical grounds has been dedicated to the issues of hard paradiplomacy or high politics. In contrast with such trends of the past, which in their narrative divide policy areas sharply, this book has attempted to identify the influence of sub-state actors by laying a bridge between the inductive (situational) and the deductive (nomothetic) approaches. In other words, in this book we endeavored to, not simply take the landscape created by the center (the host country's central government and the U.S. as the country establishing the base) for granted, but to, by shedding light on the policy choices of sub-state actors from the level of national security, delineate the landscape beyond it. For that purpose, we argued by citing not only the example of Greenland, but also the cases of Okinawa (Japan) and Olongapo (the Philippines), where in the conditions in which the sub-state actor is exercising influence on the center in an internal political process concerning the placement of a military base, analogies with Greenland can be found – that is, the questions of vulnerability and hold-up that inform the politics regarding military bases in democracies appear there too (although the outcomes are different from Greenland).

2 The Correlation between Vulnerability and Hold–Up

So, what kind of hints can be obtained in view of the overall theme of this book, the relationship between Denmark and Greenland regarding U.S. military bases, when we take into account these reference points? Let us first briefly summarize the facts concerning the relationship. In April 1941, during World War II, Denmark allowed the U.S. military to enter its colony of Greenland in order to counter the threat posed by Germany. After the end of the war, as the United States recognized the importance of Greenland for its Soviet strategy, it asked the Danish government if it could continue to use its bases in Greenland. The talks were difficult, but in the end, the two parties signed the Defense of Greenland Agreement in April 1951 (hereinafter, the 1951 agreement) in which Greenland (and within it, Thule Air Base) was redefined as an important part of the West's nuclear strategy (Petersen 1981, 1982, 1998).

Alongside this development, the relationship between Denmark and Greenland changed. In 1953, the Danish government incorporated Greenland as its an administrative region, and in 1979 granted it the legal status of a self-governing territory (the right to autonomy in domestic affairs) through the Home Rule Act. Moreover, in 2003, the Danish government concluded the Itilleq Declaration that gave Greenland certain rights in terms of having its own say in the fields of diplomacy and security, traditionally viewed as exclusive prerogatives of the state. In addition, in the following year, Copenhagen signed the Igaliku Agreement that amended and supplemented the previous 1951 agreement and positioned Greenland as an "equal partner.[1]" Thus, Greenland, as a sub-state actor, acquired the capacity of an official participant in the negotiations conducted with the United States regarding the Thule base.

So, how did the domestic political dynamics regarding military bases that we have examined in this volume function behind this expansion of Greenland's autonomy rights? First, the strategic value of U.S. military bases in Greenland, and particularly the Thule base, remained high throughout the entire Cold War period. (Henriksen and Rahbek-Clemmensen 2017) Moreover, unlike the case of Okinawa or Olongapo, Thule was practically the only security asset that the Danish government could provide to the United States (or to NATO in exchange for a membership in that alliance). (Duke 1989) In return for providing Thule, Denmark was given the status of a NATO founding member state, and at the same time became, it can be said, dependent on Greenland for its national security. (Kawana 2012)

[1] However, even in the Igaliku Agreement, the United States did not change the article on the free use of the bases, which was the core of the 1951 Agreement (2.3 (b)), thus protecting its core interests: "Without prejudice to the sovereignty of the Kingdom of Denmark over such defense area and the natural right of the competent Danish authorities to free movement everywhere in Greenland, the Government of the United States of America, without compensation to the Government of the Kingdom of Denmark, shall be entitled within such defense area and the air spaces and waters adjacent thereto." (*Agreement between the United States of America and Denmark*, Amending and Supplementing the Agreement of April 27, 1951, Signed at Igaliku August 6, 2004.) https://www.state.gov/documents/organization/170358.pdf (Last viewed on September 30, 2017).

Furthermore, the changes in the relationship between Greenland and Denmark, with Greenland evolving from a Danish colony (up to 1953), to administrative region (1953–1979), a territory possessing autonomy in internal affairs (1979–2003), and then to a territory possessing autonomy in external affairs too (2003-present), theoretically enhanced the residual control rights of Greenland. It goes without saying that this should have increased the risk of hold-up for the government in Copenhagen. The fact that the Danish government did not have the option to distribute transactions regarding military bases among political actors other than Greenland (the 1951 Agreement specified that bases could only be established in Greenland) seems to have made it difficult for it to control its vulnerability. While attention should be given to the fact that institutionalization carried out by the Copenhagen government through side payments to the government of Greenland (e.g., subsidies and other financial measures), appears to have worked, it is also worth noting that had this method been used alone (not in combination with integration or distribution), its effects would have likely been limited. In the cases of Okinawa and Olongapo, which we took up as reference axes, the option of distributionexisted. Concretely, since in the case of Okinawa no geographical limitations were placed on the location of the bases in the status agreement with the U.S. (Article 3) and the option of distributionexisted systemically, the hold-up effect was not as large as in the case of Greenland. The same can be said for Olongapo - since the geographical scope in which U.S. bases could be placed was not specified, systemically, the option of distributionexisted, which is why the hold-up effect did not function well.

What emerges as clear when we refer to the cases of Okinawa and the Philippines is that the existence of the systemic option of distribution and the increase or decrease in the possibility that bargaining regarding a military base will be transferred to a different actor (that it will be dispersed) are not automatically connected. For example, Subic Bay Naval Base, discussed in the chapter on Olongapo, illustrates this point well: it was not only the largest U.S. naval base in Asia, deemed by the U.S. as having an utmost strategic military value, but was also recognized for its importance as a logistic center for supply and repair, and thus viewed as irreplaceable. In such circumstances, relocating the base through bargaining with others become difficult and the government was not able to deal resolutely with the hold-up by the sub-state actor.

When we try to analyze the subtleties of political dynamics surrounding military bases from the standpoint of the correlation between vulnerability and hold-up, the influence of subnational entities emerges before us as a factor that cannot be disregarded. For example, the case of the agreement concerning the return of Futenma Base in Okinawa illustrates that vividly. There, spurred by the rape of a girl by an American marine in 1995, multiple meetings between the central government (the cabinet headed by Prime Minister Ryutaro Hashimoto) and the sub-state actor (Okinawan government headed by Governor Masahide Ota) took place in an attempt to both achieve sufficient strategic military deployment of U.S. troops and a reduction in Okinawa's burden of hosting the U.S. forces. On the discussion table were issues such as the treatment of the bases and the U.S. Marine Corps, as well as reallocations, and as a result of the talks, the return of the base to Japan was agreed on (Yara et al. 2016). If we turn our attention to this process, we can see that the question of the military posture of the U.S. forces in Okinawa (their numerical strength and functioning) is not just a result of the consideration of the trends in the wider security environment in Asia-Pacific, but that it is also an expression of achievements in the bargaining with a sub-state actor, as well as between states. In other words, it can be said that it is also a manifestation of political judgements by two sides, informed by a fine balance between the vulnerability and hold-up. However, it should be noted that the agreement regarding the return of Futenma Base achieved as described above, has, in fact, not been realized up to this day. That the gap between decisions and their execution is a major political issue for sub-state actors is also visible in the case of the Igaliku Agreement in Thule.

2.1 Triadic Relationship

In relation to the case of Thule Air Base, worthy of our interest is the fact that since Greenland's legal status was upgraded from "administrative region" to "autonomous territory," the actors conducting the base politics there changed from two parties (the United States and Denmark) to three parties (the United States, Denmark, and Greenland). According to past studies, because of this, signs of change appeared in the relationship between the Danish government, which had previously been "monolithic" in its approach, and the United States. Furthermore, exchanges between the United States and Greenland, which had not taken place up to that time, began from the lifting of the ban on the employment of local people in the bases (1986) (Archer 2003). Moreover, information related to the security of Greenland is now being provided by the Danish government to the Greenland Home Rule Government. In other words, to a certain extent, it has become necessary to coordinate the interests of the three parties when planning and managing base policy.[2] This occurred concurrently with the increased importance of Thule as a C3I (command, control, communication, and information) base, and particularly since 2002, when the United States withdrew from the ABM Treaty, Thule has been positioned as the core of its global missile defense system. (Taagholt and Hansen 2001)

On the other hand, Greenland strongly opposed the incorporation of Thule into the United States' missile defense system and announced that it did not approve of the strengthening of the base's functions (i.e., radar improvements) (Archer 2003: 141). This is because it did not want to create tension with Russia, and also because it perceived the missile defense concept to be contrary to the spirit of the 1951 Agreement, which was a defense-oriented agreement, as the concept was inherently offensive in nature. Moreover, in Parliament, criticism of the 1951 Agreement as a remnant of colonial era inequality grew, and its amendment was demanded.

[2] In 1995, a consultation committee on diplomacy and security between the Danish government and the Greenland Home Rule Government was established.

Therefore, the outcomes of its home rule (the expansion of its residual control rights), the dynamism of base politics, and the change from a dyad to a triad, can be said to have greatly shaken the stability of the bases in Greenland.

For the Danish administration of Anders Fogh Rasmussen that prioritized its security relationship with the United States, securing Thule's political stability and continuing the base agreement were critical issues. In other words, it seems that it was inevitable that this administration, which was being held "hostage" by Thule, would allow the further expansion of autonomy rights that were being demanded by Greenland, which may be interpreted as its show of "loyalty" to the United States. Therefore, both parties agreed to the Itilleq Declaration in 2003 and then concluded the Igaliku Agreement with the United States in the following year. This effectively incorporated Thule into the United States' missile defense shield. Of course, since then the extent of the acquisition of autonomy and its application by Greenland have hardly changed at all and are continuing to be in a dysfunctional state, as we have seen in Chapter "Greenland's Quest for Autonomy and the Political Dynamics Surrounding the Thule Air Base". Needless to repeat, this is because the intent of Greenland to use the Igaliku agreement framework to build a multilateral relationship while further enhancing its autonomy is not proportional with the extent of the issues it can deal with in reality. However, if that is so, then the perspective that focuses on vulnerability and hold-up in order to comprehend the bargaining concerning military bases as political assets between the central government and the sub-state actor, it must be pointed out, is effective since it does not simply offer an isolated snapshot view of the base, but traces historically the spatial and temporal axes of the dynamics of base politics.

3 Autonomy and the Dynamics of Base Politics

None of the authors in this volume *except* Takahashi specialize in so called "Arctic studies". What is more, none of the authors, *including* Takahashi, come from any of the coastal countries of the Arctic Ocean, or the circle north of the 45th parallel, which Laurence Smith calls "the new North". While we leave the decision on whether our positionality as outsiders has affected the content of the book in a positive or negative way to the reader, here, in response to the main theme of this volume, the influence of sub-state actors on national security, we wish to close this book not just by looking at the correlation between vulnerability and hold-up, but also by bringing up the below three points regarding the scope of autonomy indicated in the introductory chapter. These three points of argument are the essence that has functioned as, so to speak, the leitmotif of this book, which, we wish to point out, looks closely from the level of national security not just at the political bargaining between states, but also at policy choices made by sub-state actors. We believe that putting spotlight on security relations in the Arctic from such a perspective can help shed light on various political aspects of the relationship between sub-state actors and national security that is not based simply on the old, conventional logic of inclusion and exclusion in which the state is the only subject. We also think that by referring to the cases of Okinawa and Olongapo, that have been incorporated into this volume as supplementary points of reference, the below three points can be understood even more clearly.

3.1 The Relations between State Constituents and Negotiators at the Diplomatic Level

In this book, while taking up diplomatic and military affairs under the jurisdiction of the state as examples, we have focused our attention on the policy choices of sub-state actors in more detail than on the bargaining between states. When we look closely at the process in which Greenland has sought autonomy, taking the incorporation of Thule Air Base in the U.S. missile defense shield as a study case, we can see that Greenland's motives for demanding more powers have not been limited in scope just to Denmark, but also included its asymmetrical power relationship with the United States.

Concretely, Greenland encountered numerous problems regarding that U.S. base, the construction of which was prompted by the Second World War, and when after the end of the Cold War important diplomatic documents were released, started mulling policies for the dissolution of its status as a ruled entity. What it demanded then was a framework that would enable it to negotiate directly with the U.S. so that it could fulfill its duties and responsibilities as a constituent of the Danish state, and not just simply keep agreeing or disagreeing with the decisions made by the Danish state in the sphere of foreign relations. In other words, Greenland's distinctiveness can be found in the fact that, rather than simply questioning from the standpoint of an internal constituent at what level an agreement reached in international negotiations should be accepted, it sought to get involved in the international negotiations regarding the base as one of the negotiators. In past studies, however, a strong tendency existed to view internal state constituents as entities that simply question at what level agreement reached in international negotiations should or should not be accepted – a cognitive frame of reference that can be termed top-to-bottom transmission was common.

With the above in mind, we may say that Greenland's choice to not use the right it acquired to have its voice heard to draw a legal line between itself and the Danish state, but to, exactly because it has responsibilities as a subject constituting the Danish state, use it as a foothold for becoming a more self-reliant subject within the framework of that state, is an important aspect of its aspirations for autonomy. What is evident is that, although it has acquired the right to negotiate with other states and international organizations, Greenland kept the protection of the general interests of the Danish state as the premise of its actions and did not aim to affect or limit Denmark's constitutional responsibilities and authority in international relations. In a report by the Committee for Autonomy published in 2008 it is stated that Greenland should in its relations with foreign countries aspire for "a maximum enhancement of the right to self-rule for the people of Greenland within the existing state system". The wording "within the existing state system" is practically synonymous with saying "within the framework of the Danish constitution". Thus, we may understand Greenland's foreign relations as activities that are aimed at acquiring and enhancing own powers, without the intention to shake the Danish state.

3.2 *The Layer of Local–Local Relationship*

In Chapter "Base Politics and the Hold-up Problem" Kawana aptly pointed out the problem with Andrew Yeo's concept of security consensus. The problem is that, with that concept, it cannot be explained what it is that in unitary host countries makes some bases become a political issue and others not, and what the difference is between those local governments that can influence the central government and those that cannot. As an analytical perspective for overcoming that problem, in our book we have placed focus on the conditions in which a sub-state actor exercises its influence on the central government, that is, on vulnerability and hold-up as elements in the bargaining between the central government and sub-state actors regarding military bases as political trade assets.

Of course, when the U.S. as the country that establishes bases is added, a trilateral relationship, emerges but the point we wish to make through our argument in this book is that, from the viewpoint of the study of autonomy, the dynamism of the relationship of sub-state actors with other sub-state actors (local-local relationship), in this case between Nuuk and Qaanaaq, should also be taken into account as an explanatory variable that constitutes the politics of military bases. Chapter "Base Politics and the Hold-up Problem" focuses on the actors, who representing different opinions and positions, conduct the negotiations (act as the windows for negotiations), but that of course does not mean that attention has not been paid to the inherent complexities of the actors. Because, needless to say, neither the area hosting the base, nor the country installing the base, or the host country's central government can be monolithic in their respective views regarding the establishing and operation of bases.

However, considering the burden of having military bases, the problem of whether the sub-state entities, where the lives and livelihood of the locals are closely intertwined with national security, should be spoken of with the same level of tension as the central governments of the base-providing and base-receiving countries should be examined. As Hiromori Maedomari points out, in the discourse of central governments, national security tends to be referred to in terms of "contingency" or "emergency", whereas for sub-state entities national security tends to be an issue that clearly exists in peacetime (Maedomari 2015). If so, then how national security overlaps with the everyday life (peacetime) of a lower-level entity within a sub-national entity is an explanatory variable that carries more weight than the disputes within the central governments of the base-providing and base-receiving countries. The consideration of Nuuk for Qaanaaq in the matter of the transformation of Thule Air Base into a missile defense outpost, described in detail in Chapter "Greenland's Quest for Autonomy and the Political Dynamics Surrounding the Thule Air Base", is a result of Nuuk's political judgment, which took into account such internal gradation within a sub-national entity. Nuuk, argues this book, aimed for a redefinition of the relationship between the concerned political subjects (i.e., for the renewal of the defense agreement) by skillfully expressing in the negotiation process the metamorphosis in the positionality of Greenland that arose from the inclusion of that local-local relationship.

3.3 The Emergence of Cold–War–Like Power Relations in the Post-Cold War Era and their Influence on Sub-state Actors

The contribution of this book could be summarized from a somewhat more meta angle. In the past social scientists such as Martin Pratt and others warned that we should be cautious about interpretations that explain the political environment in the Arctic in terms of the new Cold War (Pratt 2010: 33). They pointed out that the differences among the coastal states of the Arctic Ocean in their views regarding sovereignty and jurisdiction in that maritime area might be resolved in the future by peaceful means. However, to what extent is such an interpretation justified when we look at the differing perceptions of threats and understandings of (maritime) order in the region today? For example, as this book has shown, in recent years in the Arctic maritime area (especially, in the waters of the North Atlantic Ocean) military activities, such as the dispatching and deployment of submarines and the building and refurbishment of naval bases by Russia, have been intensifying markedly. Also, in what may seem as a response to that, reshuffles and increases in the U.S. defense budget aimed at countering Russia have been observed. This, in effect, means that the budget intended for guarantying Europe's security (the European Reassurance Initiative) and the coordination between the United States and Europe have been reorganized and enhanced,[3] boosting the North Atlantic alliance. Of course, such activities by Russia and the U.S. will not instantly lead to an armed conflict. Nonetheless, it is undeniable that in the Arctic waters of today a "security dilemma" is visible, embodied in the race by the two sides, motivated by concerns regarding militarization of the region, to strengthen their own military assets and defense capabilities.

This security situation unfolding before our eyes indeed has a Cold War-like appearance. The above-mentioned security dilemma is not only creating tensions, but also providing a trigger for the securitization (the problematization of security) of the Arctic, as citizens grow more aware of it. Thus, posing questions about sub-

[3] US Department of Defense Budget FY2017 (http://comptroller.defense.gov/Portals/45/Documents/defbudget/fy2017/FY2017_ERI_J-Book.pdf)

jects of national security, which in conventional logic tend to be exclusively nation states, such as "Whose security is it?", and about objects of national security, such as "What is being secured?", using the yardstick of a sub-state actor who is affected by national security, is significant not only academically, but, it can be said, may also present a broader social contribution of this book mentioned in the introductory chapter. At the same time, the fact that the security environment of recent years is affected not only by threats in the form of the acquisition of hegemony over territories, but also by acts of social displacement and dispositioning by various actors, such as seen in the war on terror, possibly raises the social value of this volume which examines national (state) security from the perspective of a non-state actor. That is, we believe, that this book has the potential to help readers understand various political aspects of the two-way relationship between sub-state actors and national security that may develop in the Arctic in the future.

References

Archer, C. 2003. Greenland, US bases and missile defense: New two-level negotiations? *Cooperation and Conflict* 38 (2): 136–138.

Duke, S. 1989. *United States military forces and installations in Europe*. Oxford/New York: Oxford University Press.

Heininen, L. et al. 2014. *Security and sovereignty in the North Atlantic*. New York: Palgrave Macmillan.

Henriksen, A. et al. 2017. The Greenland card: Prospects for and barriers to Danish Arctic diplomacy in Washington. In *Danish foreign policy yearbook*, 75–98. Copenhagen: Danish Institute for International Studies.

Kawana, S. 2012. *Kichi no Seijigaku: Sengo-Beikoku no Kaigaikichikakudaiseisaku no Kigen [Base politics: The origins of the United States' post-war overseas bases expansion policy]*. Tokyo: Hakuto-Shobo.

Kristensen, K. et al. 2018. *Greenland and the international politics of a changing Arctic: Postcolonial paradiplomacy between high and low politics*. London/New York: Routledge.

Maedomari, H. 2015. Anpo womeguru nihon to okinawa no soukoku. *Okinawa gatou nihon no anzenhosyou* [Conflict regarding security between Japan and Okinawa. In *Series on Japan's national security, No.4: Okinawa questioning Japan's national security*]. Tokyo: Iwanami Shoten.

Petersen, N. 1981. The Alliance policies of the smaller NATO countries. In *NATO after thirty years*, 83–106. Wilmington: Scholarly Resources.

———. 1982. Britain, Scandinavia, and the North Atlantic treaty 1948-49. *Review of International Studies* 8 (4): 251–268.

———. 1998. Negotiating the 1951 Greenland defense agreement: Theoretical and empirical aspects. *Scandinavian Political Studies* 21 (1): 1–28.

Pratt, M. 2010. The Scholar-Practitioner Interface in Boundary Studies. *Eurasia Border Review*. Vol. 1 (1). Sapporo: Global COE Program "Reshaping Japan's Border Studies" Slavic Research Center at Hokkaido University: 33.

Taagholt, J. et al. 2001. *Greenland: Security perspectives* (trans: Lufkin, D.). Fairbanks: Arctic Research Consortium of the United States.

Yara, T. et al. 2016. *Okinawa to kaiheitai: Cyuuryuu no rekishitekitenkai [Okinawa and the U.S. marine corps: The historical development of stationing]*. Tokyo: Junposha.

MIX
Papier aus verantwortungsvollen Quellen
Paper from responsible sources
FSC® C105338

Printed by Books on Demand, Germany

MIX
Papier aus verantwortungsvollen Quellen
Paper from responsible sources
FSC® C105338

Printed by Books on Demand, Germany